MINING METHODS

Edited by **Turgay Onargan**

Mining Methods
Edited by Turgay Onargan

Published by InTech

Janeza Trdine 9, 51000 Rijeka, Croatia

Publishing Process Manager Petra Nenadic

Technical Editor Teodora Smiljanic

Cover Designer InTech Design Team

Additional hard copies can be obtained from orders@intechopen.com

Mining Methods, Edited by Turgay Onargan
p. cm.
ISBN 978-953-51-0289-2

For EU product safety concerns:
IN TECH d.o.o., Prolaz Marije Krucifikse Kozulić 3, 51000 Rijeka, Croatia,
info@intechopen.com or visit our website at intechopen.com.

Meet the editor

Dr. Turgay Onargan is a mining engineer who has specialized in rosk mechanics and mine production methods. He hold a BS in mining engineering and PhD in mining engineering. Before obtaining his Doctorate in the field of underground openings design, he spent four years in mine planning at the Turkish Coal Enterprises from 1986 to 1989. In 1989, he joined to Department of Mining of Dokuz Eylul University in Izmir, Turkey. His expertise includes rock mechanics, mining methods, tunneling, mine surveying, strata control and natural stone mining. He has won awards for his research. His consulting activities involve mine planning and rock mechanics (including tunneling) in Turkey.

Contents

Preface

Recovery of minerals from subsurface rock involves the development of physical access to the mineralized zone, liberation of the ore from the enclosing host rock and transport of this material to the mine surface. Excavations of various shapes, sizes, orientations and functions are required to support the series of operations which comprise the complete mining process.

Several methodologies have been developed in the past to evaluate suitable mining methods for an ore deposit, based on the physical and mechanical characteristics of the deposit such as shape, grade, and geo-mechanical properties of the rock. Mining method selection is the fundamental decision made in a mine project, and a proper choice is critical as it affects almost all other major decisions. Also, the selection of a suitable mining method for an ore deposit involves consideration of a diverse set of criteria.

This book describes methods for the surface and underground mining of mineral deposits. The descriptions are generalized and focus on typical applications from different mining areas around the world, keeping in mind, however, that every mineral deposit, with its geology, grade, shape, and volume, is unique.

The book was divided into two main sections. The first section is about the surface mining and provides examples of coal mining applications from Australia and China. The second section covers the underground mining applications from Turkey and Poland. This book will help identification of useful concepts and ideas in the light of achievements made to date and strengthen research effort for the benefit of mining industries.

Prof. Dr. Turgay Onargan
Dept. of Mining Engineering
Dokuz Eylul University
Izmir
Turkey

Part 1

Surface Mining

1

Surface Coal Mining Methods in Australia

Rudrajit Mitra and Serkan Saydam
The School of Mining Engineering,
The University of New South Wales, Sydney,
Australia

1. Introduction

Minerals are one element of Australia's natural resource base. Other key natural resources include forestry and fisheries, and together with agriculture, they make a significant contribution to the Australian economy and Australia's Gross Domestic Product (GDP). Mining contributed 7.7% Gross Value Added in 2008-2009, up from 4.9% in 2005-2006 (University of Technology Sydney [UTS] and Monash University, 2010; Australian Bureau of Agricultural and Resource Economics [ABARE], 2010).

Australia's export earnings from energy and mineral commodities in 2010 were $A165 billion, 25 per cent higher than in 2009 (ABARE, 2010). Australia's black coal exports were worth almost $A43 billion in 2010, second only to iron ore ($A49 billion) which was Australia's biggest export earner for that year. The total production of raw black coal in Australia in financial year 2010 - 2011 was 405 million tonnes (Mt). Black coal occurs in all States of Australia and the Northern Territory, but New South Wales (NSW) with 46% and Queensland (QLD) with 38% have the largest share of Australia's total identified resources. Queensland (55%) and NSW (42%) produce the most black coal with locally significant operations in Western Australia, South Australia and Tasmania. Figure 1 shows the black coal locations in Australia.

Figure 2 indicates the Australian electricity generation by fuel in 2008/2009. The major uses of Australian coal are

- electricity generation - thermal or steaming coal
- steel industry - coking coal.

In Australia, the major coal mining methods are as follows:

- Underground mining operations:
    - Longwall mining, and
    - Bord and pillar mining (also known as room and pillar mining).
- Surface mining operations
    - Open cut mining (strip mining)
        - Using draglines,
        - Using truck and shovels, and
        - Using a combination of above (integrated mining systems).

- Highwall mining
    - Continuous highwall mining, and
    - Auger mining.

This chapter will only focus on surface mining operations in the Australian coal industry.

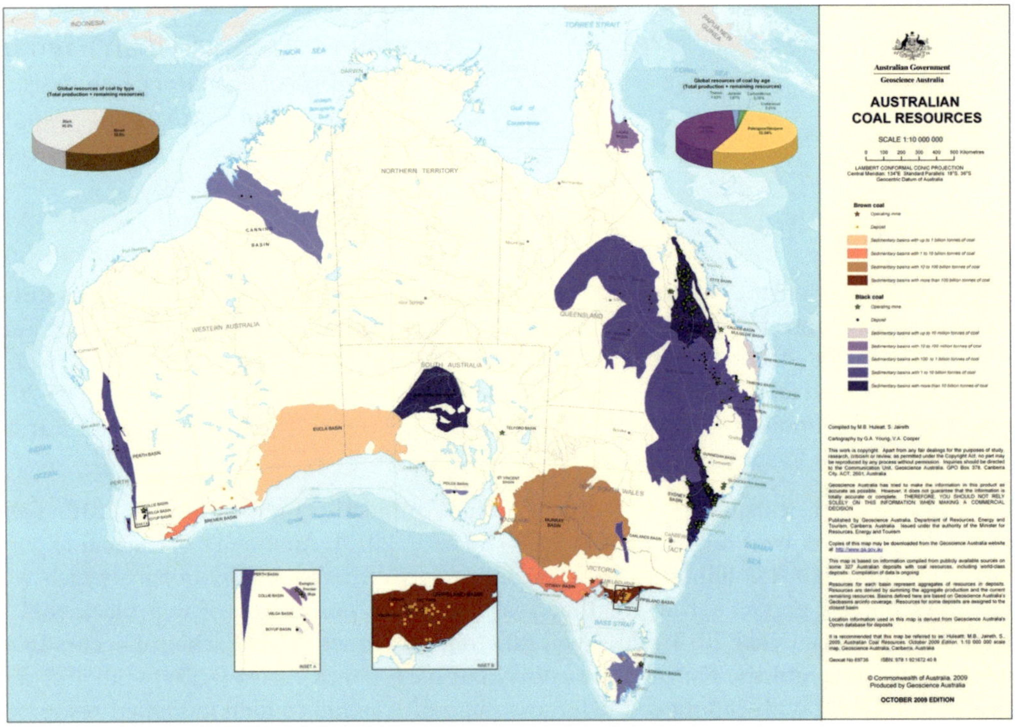

Fig. 1. Australian coal resource map (Geoscience Australia, 2009).

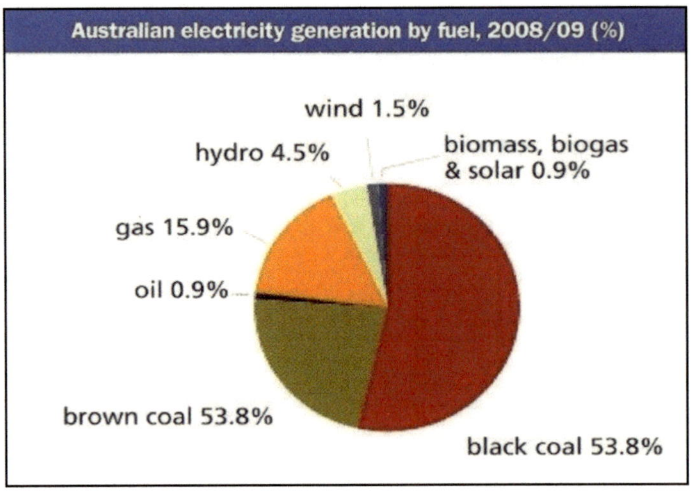

Fig. 2. Electricity generation by fuel (ABARE, 2010).

2. Open cut mining methods – strip mining

Large-scale open cut coal mining operations commenced in Australia in mid 1960s and since then there has been significant developments in this method of mining. The mines are now operating at significantly higher annual tonnages, growing deeper, more complex and operating at higher stripping ratios (Westcott et al., 2009).

The following major equipments are used in Australian coal mines:

- Draglines,
- Trucks,
- Shovels
- Bucket wheel excavators,
- Crushers and conveyors,
- Scrapers,
- Dozers,
- Slushers and dragline hoppers, and
- Surface continuous miners.

Initially open cut mines in coal mining was classified as strip mines where draglines were used, and open cut mines where truck and shovels were used. Currently a lot of mines use both equipment as integrated systems in Australia. Strip mining is ideally applied where the surface of the ground and the ore body itself are relatively horizontal and shallow and where a wide area is available to be mined in a series of strips (Westcott, 2004). Table 2 provides some examples of the Australian surface coal mines. Figure 3 illustrates a typical overall mining sequence of strip mine operation.

Mining method description	Example mines
Multiple seam, steep dip, truck and loader, strike cut haulback mine with truck and loader coal mining.	Mt. Owen (NSW), Leigh Creek (South Australia), Duralie (NSW), Ravensworth East (NSW)
Several seam, moderate dip, dragline, down dip strip mine with throw blast, dozer assist and truck and loader prestrip.	Peak Downs (QLD), Norwich Park (QLD), Hail Creek (QLD), Curragh (QLD), Moura (QLD), Hunter Valley - Riverside (NSW)
Multiple seam, moderate dip, truck and loader, block cut, haulback mine.	Liddell (NSW), Wambo (NSW), Camberwell (NSW), Hunter Valley - Cheshunt (NSW)
Multiple seam, moderate dip, truck and loader, down dip strip, haulback mine.	Mt Arthur North (NSW), Muswellbrook Coal (NSW)
Thick, single seam, moderate dip, dragline - down dip - strip mine with throw blast, dozer assist and truck and loader prestrip.	Blair Athol (NSW), Newlands (QLD), Gregory (QLD), Goonyella/Riverside (QLD), Saraji (QLD)
Thick, single seam, moderate dip, bucketwheel excavator, block, cutback mine.	Yallourn (Victoria), Yoy Lang (Victoria), Hazlewood (Victoria)
Multiple seam, moderate dip, dragline down dip, strip mine with throw blast, dozer assist and truck and loader prestrip	Bengalla, Drayton, Bulga, Warkworth, Hunter Valley - Howick

Table 1. Examples of Australian surface coal mines (adopted from Westcott et al., 2009).

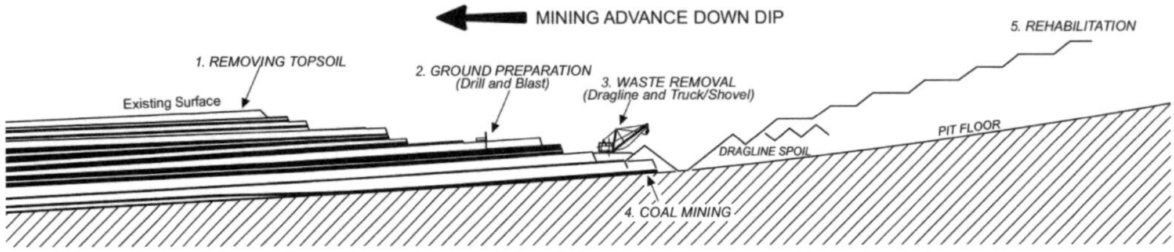

Fig. 3. Overall mining sequence of a typical surface coal mine in Australia (after Westcott et al., 2009).

2.1 Dragline method

Dragline is the predominant machine which is used to remove the overburden and expose the coal when the deposit's characteristics match the draglines' physical capabilities. Once the characteristics of the deposit alter from the physical limitations of the dragline, the overburden removal becomes costly as the machines' rehandle increases (Scott et al., 2010). In this situation, typically in multi-seam mines, truck and shovels are introduced to create an integrated mining sequence, which operates in tandem to generate a cost effective and efficient mining system (Fox, 2011).

Draglines are the lowest cost overburden removal equipment in common use. However, they are generally restricted to (Westcott et al., 2009):

1. Large deposits to ensure adequate strip length and sufficient reserves to justify the capital expenditure.
2. Gently dipping deposits, due to spoil instability on steep dips.
3. Shallow deposits, as draglines can only excavate a maximum of 50 to 80 m of overburden due to reach and dump height limitations. At greater overburden thicknesses draglines may be supported by pre-strip with alternative equipment, but the compromises this entails usually adversely affects the cost effectiveness of the pre-strip.

There were 68 draglines operating in QLD and NSW in 1992, with additional draglines either being built or ordered (Baafi and Mirabediny, 1998). While the most common machines average a 46m^3 bucket size, there has been a trend in favour of larger machines. These draglines feature much larger bucket sizes (up to 115m^3) and dimensions allowing the draglines to operate at greater depths and widths without needing pre-stripping and reducing rehandle (Aspinall et al., 1993; Gianazza, 2010). A typical dragline section is shown in Figure 4.

Draglines work in strips, which are typically 40 to 90 m wide and few kilometres long and the overburden is excavated by the dragline and dumped on the surface for the initial strip (box cut) and subsequently in adjacent mined out strips as illustrated in Figure 5.

Strips are generally aligned along strike with each subsequent strip down dip from the previous strip. The dragline starts at one end (or the middle) of the strip and advances along the strip to the other end. At the completion of each strip the dragline relocates to the start of the strip and commences the next strip, this is referred to as deadheading. Ramps for coal

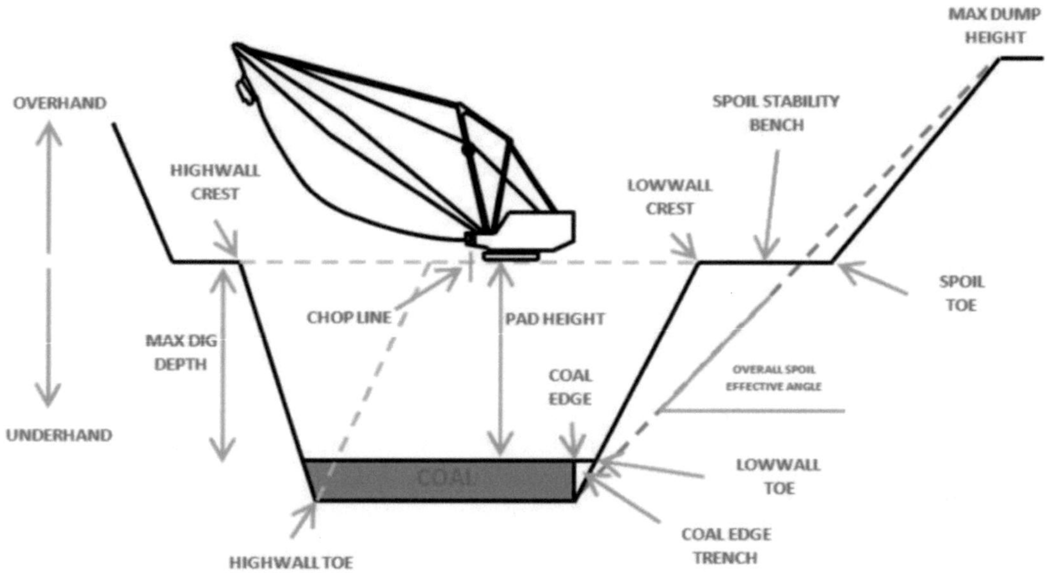

Fig. 4. Dragline Working Section (Gianazza, 2010).

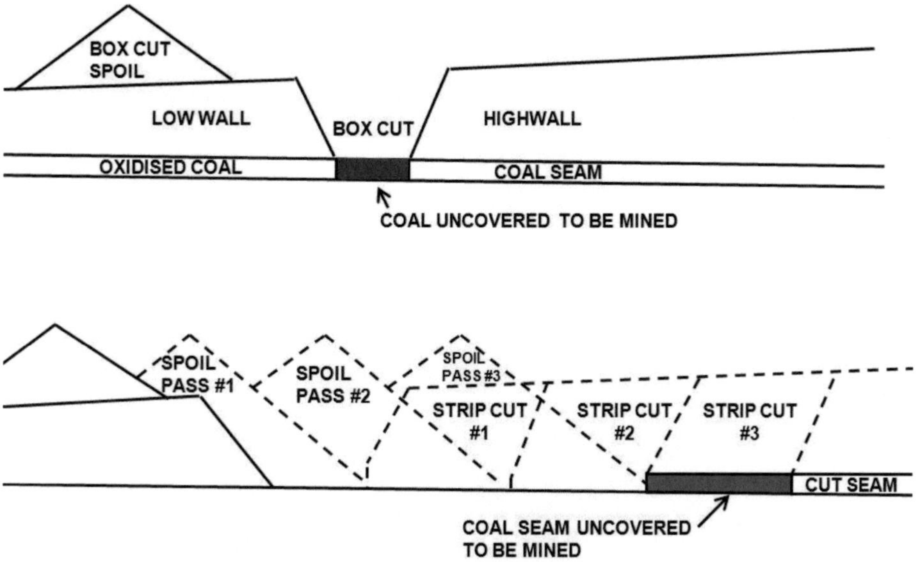

Fig. 5. Box cut, strip cuts and spoil piles (Humphrey, 1984).

mining access are either taken through the overburden dump as a valley, dozed through the spoil pile parallel to the strip, created in the highwall either parallel or at right angles to the strip, or by a combination of these alternatives. Draglines are often supplemented by throw blasting and/or dozer assistance to increase overburden removal capacity (Westcott et al.,

2009). Figure 6 illustrates a typical dragline while removing the material in one of the NSW mines.

The following are some of the advantages of using a dragline:

- Direct cast (excavate and transport),
- Low operating cost, and
- Handles hard digging.

The following are some of the disadvantages of a dragline:

- Constraints on dig depth and dump height,
- Relatively inflexible,
- Requires detailed planning, and
- High capital cost.

The rigging on a conventional dragline weighs 20 tonnes and this has not changed in design in 100 years. This conventional rigging system limits the flexibility of the dragline operation and makes bucket control difficult. Along with high productivity costs when a dragline is out of action for rigging maintenance, there is also substantial safety hazards associated with replacing heavy pieces of equipment.

Fig. 6. A typical dragline in operation in Hunter Valley Region in NSW.

A new technology was developed by the Cooperative Research Centre for Mining (CRCMining) Australia called the Universal Dig and Dump system (UDD). This replaces conventional rigging with a lighter, innovative configuration, improving operational flexibility. A medium sized UDD can move up to 13 tonnes more dirt in each pass and a specially designed computer system provides precise control over the bucket, enabling the dragline to dig and dump anywhere under the boom (Mining Technology Australia, 2011). This was developed in conjunction with BHP Billiton Mitsubishi Alliance (BMA). This has resulted in a significant improvement in productivity and is being installed on many draglines operating in Australia. Figure 7 shows one of these buckets in operation.

Fig. 7. A UDD bucket in use (Mining Technology Australia, 2011).

2.2 Truck and shovel method

Truck and shovel mining method is the most flexible mining method and therefore better suited to geological complex deposits, varying overburden depths and thicknesses, and smaller deposits (Westcott, 2004). They offer cheaper capital investments than draglines, however they cost more to operate on a bcm (bank cubic metre) per hour removal rate. Another major benefit with the system is the ability to haul material long distance to ensure rehandle will not become an issue in the present time and future of the operation (Fox, 2011).

Truck and shovel method is inherently more flexible than dragline methods which makes them better suited in the following applications (Westcott et al., 2009):

1. Geologically complex deposits with resultant irregular pit shapes, which could not be efficiently mined by a dragline.
2. Steeply dipping deposits, where the equipment cannot operate on the seam roof and floor. Mining commences at one end of the deposit and advances along strike with strips laid out down dip. Overburden is initially dumped expit and then inpit when sufficient dump room is available. The pit is excavated as a series of horizontal benches (terraces) and coal and waste are exposed on every bench. Each bench extends from the floor of the lowest seam to the down dip economic pit limit.
3. Basin deposits that combine the problems of steep dips at the margins with short strike length and varying overburden depth along the strip.
4. Small deposits, which do not require the high productivities gained through use of a dragline.

'Haulback' mining is a particular application of a loader and truck operation. It refers to the fact that the waste is 'hauled' (transported) and dumped 'back' into the previously mined out area. Where the advance is down dip the configuration is not that different from a dragline strip mine and techniques such as throw blast and doze assist can also be incorporated. A good example of this design is Mount Arthur North of BHP Billiton in the Hunter Valley, NSW. To keep haul distances as short as possible, many operations make use of rehandle bridges across the strip to transport waste from lower benches to the dump. As mining depth increases, a greater proportion of waste is long hauled around pit endwalls to the upper dump levels. 'Terrace mining' also dumps inpit but mines along the strike so the waste dump is keyed into a highwall. Multiple waste haulroads link the mining terraces to waste benches. This minimises haul cycle times for waste removal by reducing the vertical component of the haul. For the technique to work successfully both the mining terraces and the dump benches must be offset from adjacent benches to allow operational scheduling of mining activities. The method relies on the highwall being planned in its final position because once the mine is backfilled it would be a big economic hurdle to rehandle the waste to add another down dip increment. 'Cutback' open pit mining is where the waste is not dumped inpit but the waste is hauled to an external dump. Reasons for not dumping inpit could be that the floor is too steep for stable dumps or else the option of coming back later to either deepen the pit or extend the highwall down dip are part of the long-term plan. Panel shapes and sizes are dictated by production requirements, available equipment fleet and deposit geometry, rather than by parallel strips. Cutback mining may be used equally well in steeply or shallowly dipping deposits (Westcott et al., 2009).

The versatility of the system and ability to haul large distance makes this system extremely favourable in nearly all mining situations (Hays, 1990). Once equipment has been selected and the four main issues of cycle time are targeted and managed as best as possible to the conditions, will the system perform and achieve high production rates (Fox, 2011). A typical truck and shovel operation can be seen in Figure 8.

Fig. 8. A view of a truck and shovel operation in Hunter Valley Region in NSW.

2.3 Integrated mining system – combination of truck & shovel and dragline

The integrated mining system where truck and shovel is combined with the use of a dragline for overburden removal, provides a system with many benefits. Generally, the truck and shovel system is used to remove the upper and thinner overburdens found within a deposit and the draglines remove the much deeper overburdens, which would be out of the working range of a shovel operation (Aiken and Gunnett, 1990), as shown in Figure 9.

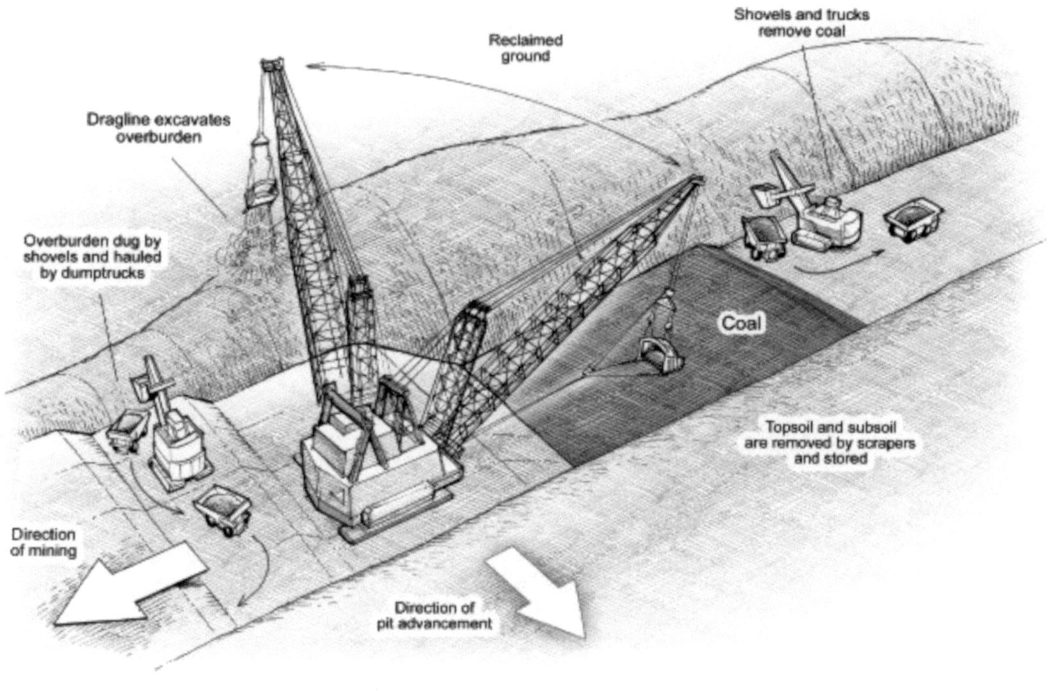

Fig. 9. Integrated Mining System - combination of Truck & Shovel and Dragline methods (after Fox, 2011).

3. Case studies

3.1 Mount Thorley Warkworth mine, NSW

Mount (Mt.) Thorley Warkworth is an open cut coal strip mine located in the Hunter Valley, approximately 15km south of Singleton, NSW. The mine is owned by Coal & Allied and is operated by Rio Tinto Coal Australia, which is one of the largest mining companies in the world. Mt. Thorley Warkworth is a joint venture between Coal & Allied Industries Limited (80%) and Posco Australia Pty Ltd (20%). Coal & Allied has been around for 150 years and along with Mt. Thorley Warkworth. Figure 10 shows the location of Mt. Thorley Warkworth mine in the Hunter Valley Region.

Fig. 10. Mt. Thorley Warkworth locality map (after Fox, 2011).

Mt. Thorley Warkworth Mine currently produces around 12mtpa (million tonne per annum) of thermal export coal with the possibility of ramping up to 15mtpa in the future. The mine has operated an integrated mining system since the site's commencement in 1981. The site has two processing plants on site, due to the merger of the pits, and the coal type and quality depend on which plant the coal is delivered to. Once the coal is washed, it is then transported with the help of a conveyor to the coal loading unit for transport to Newcastle Port for exportation. Currently, the fleet system has 76 trucks ranging from 190t to 240t trucks, with the purchase of 330t trucks scheduled for 2012 (Corrigan, 2011).

The draglines operate in the lower portion of the seams where the interburden are the greatest and the pre-strip shovels and excavators are used on the upper portion of the deposit, where the interburden average is around 15-20m. The highwalls are 70 degrees, lowwalls are 45 degrees and the spoil and dump angles are 37 degrees. These angles allow the pit to operate safely as the geotechnical conditions are extremely favourable (Fox, 2011).

Mt. Thorley Warkworth Mine has a life of mine until 2050 and with the current coal processing capacities, the site has the ability to achieve a much higher production rate and increase the revenue generated. Figure 11 illustrates the aerial view of the mine and Figure 12 shows the truck and shovel system currently operating and removing the pre-strip material.

Fig. 11. Mt. Thorley Warkworth Mine aerial view (Fox, 2011).

Fig. 12. Mining sections in Mt. Thorley Warkworth Mine (Fox, 2011).

3.2 Curragh mine, QLD

Curragh Mine is located in the Central Queensland's Bowen Basin coal fields near Blackwater and operated by Wesfarmers Curragh Pty Ltd. The first successful extraction of coal occurred during the 1960's in the area. In 1984, Curragh Mine began export of hard coking coal, mostly to Japanese and Korean smelters. Since then, the mine has undergone two major expansions: developing the Curragh East deposit on the eastern side of Blackwater Creek and the Curragh North deposit, which more than doubled the recoverable reserves available in 2005 (Gianazza, 2010). Figure 13 shows the location of the mine.

Curragh Mine currently has a mine life up to 2025, with over 220 million tonnes of coal reserve within the mine plan and produces three different coal products: export metallurgical, domestic thermal and pulverised coal injection coals, each with varying specifications and parameters. The mine has a unique combination of complex geological settings and equipment compared with other mines in the Bowen Basin, QLD. Mining operations in Curragh Mine have five major seams with four of these seams mined in a single pit. The average seam thicknesses for each of the coal seams mined vary between 1 and 2.5m at the mine. Due to the geological complexity of the mine, the process differs significantly from pit to pit. Developed over a strike length of approximately 16 km, mining at Curragh Mine commenced at the line of oxidation (LOX) and has progressed down dip to the east. There is also a general trend in the topography sloping towards the east, limiting the increase in prime stripping ratio. Prior to removal of overburden, vegetation is removed, the land is assessed for indigenous significance in the area, and the topsoil stripped and stockpiled for later use in rehabilitating previously mined areas. With the exception of the first top 20m at Curragh North, all overburden and interburden are drilled and blasted for removal through either truck and shovel or dragline operations (Gianazza, 2010).

Fig. 13. Location map of Curragh Mine (Gianazza, 2010).

Curragh Mine operates four large capacity walking draglines as primary stripping machines, including one of the world's largest, with a $114 m^3$ bucket. Depending on the pits and faulted areas, the draglines generally operate a multi-pass digging method, with strip widths usually in the range of 60-70 m range and highwalls designed at 63 degrees. The mine also operates a unique fifth dragline which provides significant operational flexibility, performing specific tasks more efficiently and effectively than a truck and shovel operation. This significantly facilitates overburden removal around highly faulted areas, due to its small size and high mobility. The coal removal process is undertaken by multiple fleets of hydraulic excavators, belly-dump coal haulers (204t) and rear dump trucks (180t). Coal is transported to the Coal Handling and Processing Plant (CHPP) and discharged at two 500t hoppers. The CHPP is dual line in design, with total crushing capacity of 2,400t of raw coal per hour. The plant allows flexibility by running coal to the preparation plant feed stockpiles through one line, and running bypass steam coal through another. Simultaneously processing two different coal types, the plant handles more than 1,400t of coal per hour through three modules. Product coal is stacked onto two stockpiles, each capable of holding 150,000t of coal. A bucket-wheel reclaimer loads the 7,000t coal trains at a rate of 3,000t per hour (Gianazza, 2010).

A total of 7 truck and shovel circuits combined with the operation of 5 draglines and the largest single span conveyor in the Southern Hemisphere, makes Curragh Mine both unique and successful at being one of QLD's leading producers of metallurgical and thermal coal (Brown, 2007). A monitoring system was placed in all 5 draglines in 2009, which provides production monitoring, high-precision GPS (Global Positioning System) and bucket depth control information back to operators.

3.3 Mount Arthur Coal mine, NSW

Mt. Arthur Coal Pty Ltd, fully-owned by BHP Billiton, operates the Mt. Arthur Coal open cut mine situated 5km south of Muswellbrook in the upper Hunter Valley of NSW (Figure 14). The mine produces thermal coal used for power generation. The coal is sold to both domestic and export markets. Production capacity of the mine is 20mtpa of raw energy coal and the operation shares the area with rural properties and other industries such as horse studs, vineyards, olive groves and residential suburbs.

Current mining activities are focused on 21 unique seams which were identified during the exploration stage. Mining is conducted using truck and shovel mining fleet. Mining equipment consists some of the largest and quietest trucks engineered in the world. The overburden mining fleet consists of rear dump trucks and rope shovels, while the coal mining fleet consists of trucks and a combination of excavators. Mining activities occur 24 hours per day, 7 days per week, on 12 hour shifts supported by approximately 1,000 full time equivalent personnel.

Mt. Arthur Coal currently operates under separate planning approvals for the Bayswater No. 3 Mine, Rail Loading Facility, Mt. Arthur North Mine, South Pit Extension, Exploration Adit and the Mt. Arthur Underground (Figure 15). Open cut mining is conducted via a multi-bench, multi-strip shovel and excavator operations, which provide for the greatest operational flexibility and efficiency in the staged recovery of the coal resource at Mt. Arthur Coal. The extraction of coal via open cut methods is currently undertaken from the Mt. Arthur North and South Pit Extension area, as well as the Bayswater No 3 area, in which Saddlers Pit is mined using a contract fleet. To date only a limited amount of underground coal has been extracted from the Woodlands Hill seam via the Exploration Adit (Bailey, 2009).

Fig. 14. The location map of Mt. Arthur Coal (Bill Jordan & Associates, 2009).

Large electric shovels are used to remove overburden and excavators are used to load coal into large capacity haul trucks for transportation to the CHPP and it is located adjacent to the mine. Figure 15 shows an example of an excavator dumping onto trucks. Preliminary crushing takes place at the base of the hoppers before a conveyor transports the coal to the crushing station. Up to 2,000t of raw coal is fed through the crusher and 1,200t through the preparation plant per hour. The CHPP is where the coal is crushed, screened, cleaned and sorted according to market requirements. After the coal has been removed, rehabilitation is a priority at the mine. The overburden is replaced, shaped, covered with topsoil and replanted to resemble the original landscape (Umwelt, 2006).

Fig. 15. Excavator removing the overburden and dumping onto trucks at Mt. Arthur Coal Mine (after Tredinnick, 2005).

Once crushed, the coal is transported to one of three locations:

- The domestic coal stockpile and then transported by conveyor to the local power station. Approximately 1.7 million tonnes of coal is used for domestic energy generation each year.
- The coal preparation plant where any impurities are removed, and then transported by conveyor to the export coal stockpile.
- Directly to the export coal stockpile and then to the export coal rail load-out facility for transportation to the Port of Newcastle. Approximately 10.7 million tonnes of coal was transported by rail to Newcastle in 2010.

The coal mining industry is concerned to minimise its impact on all aspects of the environment - visual landform, air and water quality, noise levels, native flora and fauna, soil conditions, and historic, indigenous and archeological sites (Australian Coal Association, 2011). One of the best examples of an environmental management of mining activities is Mt. Arthur Coal Mine. In 2010, approximately 12.3 million tonnes of coal was transported from Mt. Arthur Coal. From initial mine planning through to final site rehabilitation, the mine aims to minimise the impact of the operations on the natural environment and nearby residents.

A range of environmental issues and social impacts need to be managed at the mine site. The degree to which these are involved with the mine planning and operations department is determined by the location of the site and the inherent characteristics of the deposit and surrounding environment. Mt. Arthur Coal has a large number of statutory approvals relating to environmental management that cover the different activities on site. In total the operation must comply with more than 1,500 approval constraints, in addition to the requirements of government policies and legislation (Tredinnick, 2005).

The proximity of the Mt. Arthur Coal operations to the community of Muswellbrook exacerbates certain facets of the operation (Figure 16).

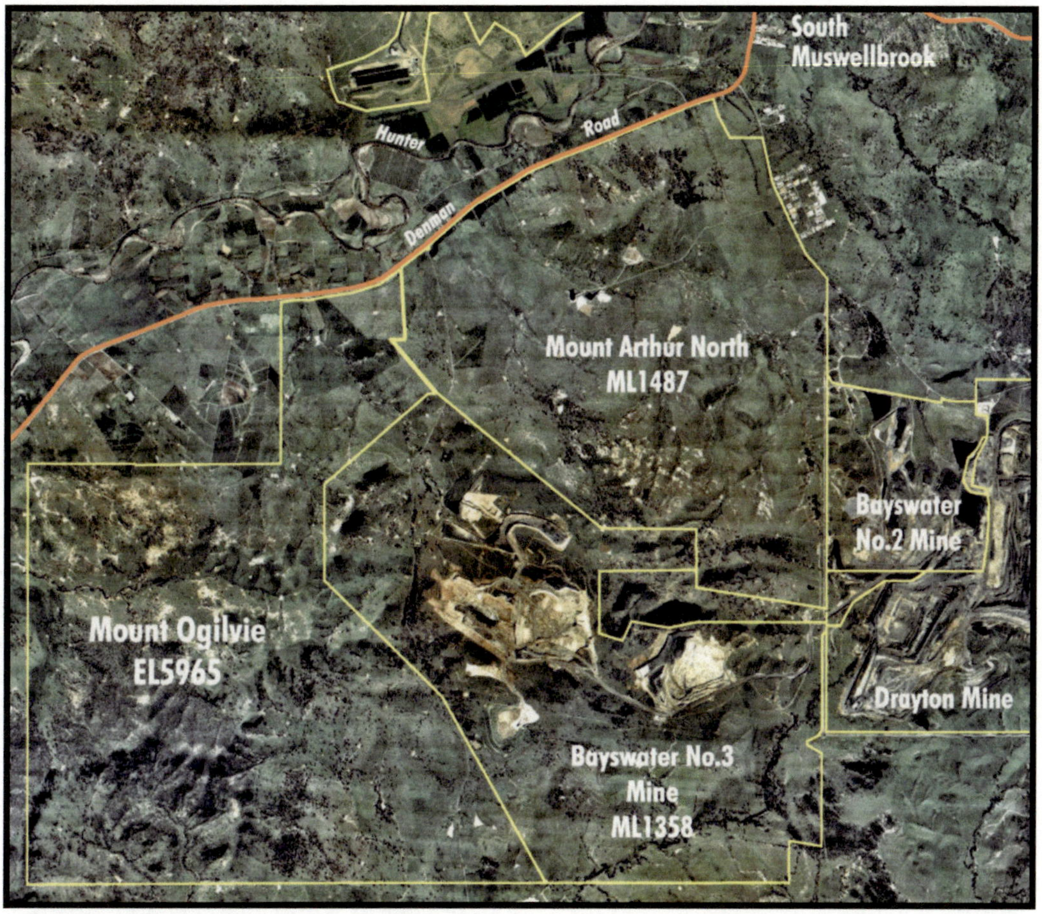

Fig. 16. Mt. Arthur Coal Mine mining lease (after Tredinnick, 2005).

Effective management of these facets involves detailed mine planning. Some key facets of the operation, from a community perspective, that need to be addressed include (Tredinnick, 2005):

- Noise: Mt. Arthur Coal is required to meet some of Australia's most stringent noise limits for mining operations. These limits are set by government and are based on noise levels considered acceptable to the surrounding community. Haul trucks, dozers and excavation equipment are the most significant source of ongoing noise. The coal handling and preparation plant, and workshop areas are further sources of operational noise, although they have been located to minimise noise impact. From the first stages of mine design a noise model was developed to allow various mine designs and fleet configurations to be evaluated. Mt. Arthur Coal purchased a fleet of trucks that are the quietest trucks in the world. These trucks produce 16 times less noise output than a standard truck.

- Visual Amenity: The mine has an extensive visual impact management by doing progressive rehabilitation. To ensure safe working on night shift there is a requirement to illuminate working faces and overburden dumps. This use of lighting plants in turn has the potential to impact on nearby residents. There is an operational requirement to plan the location of all lighting plants to ensure that safe working conditions are maintained while at the same time minimising impacts on the local community.
- Blasting: The ground vibration and over pressure from blasting is strictly regulated. The 100% limits for Mt. Arthur Coal are a ground vibration of 10mm/s and an over pressure of 120 dBL. In addition, blasting is not allowed if the wind speed exceeds 10m/s. Actual ground vibration and over pressure results, as averaged across all monitoring sites, are approximately 0.3mm/s and 100dBL. In addition to ground vibration and over pressure NO_x and dust generated by blasting are issues of concern to the local community.
- Spontaneous Combustion: Mt. Arthur Coal has committed considerable truck resources to long hauling clay rich overburden to the east pit to cap off areas of spontaneous combustion. This has seen over 95% of the spontaneous combustion eliminated.

4. Conclusions

Robertson et al. (2009) stated that the inexorable growth in world population and living standards places demands on coal resources for energy and metallurgical coke production. Within one more generation the global demand for food and energy will double. Coal is abundant and has many advantages: globally distributed, price is affordable and stable and readily converted to other valuable products. There is enough coal reserves for approximately next 120 years, which is almost 2 times more than gas and 3 times more than oil reserves. Coal provides 27% of global primary energy needs and generates 41% of the world's electricity. Australia has abundant high quality coal reserves and its geographical location close to the Asia-Pacific market has stimulated the development and expansion of the coal industry to become Australia's highest export earner. Geology, freight distances, political stability, established infrastructure and maturity enhance Australia's role in the supply of coal to world markets and Australia has vast resources of economically recoverable black coal reserves, capable of supplying demand for decades (Robertson et al., 2009). Australia is a world leader in surface mining technologies. In Australia, approximately 77% of black coal is produced from open-cut mines and surface mining equipment has increased dramatically in size over recent years. Recent developments are being made across various areas of the mining industry. With latest research and development achievements in automation, the surface coal mining productivity in Australia will continue to grow with lower cost and more environmental friendly.

5. References

Australian Bureau of Agricultural and Resource Economics [ABARE] (2010). *Energy in Australia*, Australian Government, 13.12.2011, Available from <http://www.abare.gov.au>

Australian Coal Association (2011). *Coal & The Environment - Coal Mining & Environment*, 14.12.2011, Available from <http://www.australiancoal.com.au>

Aiken, G.E. and Gunnett, J.W. (1990). Overburden removal. In: *Surface Mining*, 2nd Edition, (Ed: B A Kennedy) pp. 584-619, Baltimore: Maryland, USA.

Aspinall, T., Isles, P. and Davidson, C. (1993). Strip Mining, In: *Australasian Coal Mining Practice* (Eds: A J Hargraves and C H Martin), pp. 210-220, The Australasian Institute of Mining and Metallurgy [AusIMM]: Melbourne, Australia.

Baafi, E.Y. and Mirabediny, H. (1998). Dragline digging methods in Australian Strip Mines – A Survey, In: *Proceedings of Coal 1998: First Australasian Coal Operators' Conference* (Ed: N Aziz), pp. 313-324, University of Wollongong and the AusIMM, Wollongong, Australia.

Bailey, H. (2009). *Mt Arthur Coal Consolidation Project - Environmental Assessment*. Available from <http://majorprojects.planning.nsw.gov.au>.

Bill Jordan & Associates. (2009). *Appendix H - Blast Vibration Vulnerability Assessment*. Available from <http://majorprojects.planning.nsw.gov.au>.

Brown, N. (2007). *Post blast evaluation using dragline diggability*, Bachelor thesis (unpublished), University of Queensland, Brisbane.

Corrigan, M. (2011). *Analysis of effect of carbon tax on operational costs of the Mount Thorley Warkworth truck fleet*. BE Thesis. School of Mining Engineering, The University of New South Wales.

Fox, R. (2011). *A Study on the effectiveness of mine design software for pit optimisation and scheduling at the Mt. Thorley Warkworth Coal Mine*. BE Thesis. School of Mining Engineering, The University of New South Wales.

Geoscience Australia (2009). *Australian Coal Resources*. Available from http://www.ga.gov.au.

Gianazza R. (2010). *Strip optimization at Curragh mine*. BE Thesis. School of Mining Engineering, The University of New South Wales.

Hays, R.M. (1990). Trucks, in Surface Mining (2nd Edition), (Ed: BA Kennedy) pp. 672-691, Baltimore: Maryland, USA.

Humphrey, J .D. (1984). *The Fundamentals of the Dragline*, 29 p (Marion Division-Dresser Ind. Inc: Marion).

Mining Technology Australia. 2011. http://www.miningtechnologyaustralia.com.au/smart-technology

Robertson, B., Ford, F., Osborne, D., Seib, W., Smith, G. andVan de Ven, P. (2009). The Australian Coal Industry Future Developments. In: *Australasian Coal Mining Practice*, Chapter 45, Monograph 12, Third Edition, Eds. Kininmonth, R.J and Baafi, E.Y. pp. 1057 - 1074, AusIMM, Melbourne, Victoria, Australia.

Scott, B., Ranjith, P.G., Choi K. and Khandelwa, M. (2010). A review on existing opencast mining methods within Australia, In: *Journal of Mining Science*, 46(3): 280-297.

Tredinnick, I. (2005). *Environment & Community Opportunities and Challenges for Mine Planning and Operations*, Presented at the School of Mining Engineering, University of New South Wales, May 2005 Available from

<http://www.mining.unsw.edu.au>.

Umwelt Australia Pty Ltd (2006). Mt. Arthur Coal, Environmental Assessment Proposed Exploration Adit, Available from < htpp://www.umwelt.com.au/mtarthurcoal-adit>

University of Technology Sydney [UTS] - Institute for Sustainable Futures and Department of Civil Engineering Monash University. (2010). *Peak minerals in Australia: a review of changing impacts and benefits 2010*. Available from <http://csiro.com.au>.

Wescott, P, 2004. *The Inaugural UNSW/ Mitsubishi Lecture: Dragline or Truck and Shovel? - Some technical and business considerations*. Presented to the School of Mining Engineering, University of New South Wales, May 2004. <http://www.mining.unsw.edu.au>.

Westcott, P., Pitkin, G. and Aspinall, T. (2009). Open-cut mining. In: *Australasian Coal Mining Practice,* Chapter 18, Monograph 12, Third Edition, Eds. Kininmonth, R.J and Baafi, E.Y. pp. 410 – 458, AusIMM, Melbourne, Victoria, Australia.

2

Surface Coal Mining Methods in China

Changsheng Ji
School of Mines, China University of Mining and Technology
Xuzhou, Jiangsu,
PR China

1. Introduction

China is the largest coal producing country in the world with annual production rate of 324 Mt in the year 2010 [1]of which surface mining production is only about 9 percent of the total production because the coal reserves suitable for surface mining is not large enough compare with the other major coal producing countries. Surface coal mining methods used in China have also been limited because the geological, topographical and coal occurrences conditions.

2. Recoverable coal deposits to surface mining

2.1 Introduction

Coal reserves are available in almost every province in China, with recoverable reserves in around 334.1billion tons[2]. At current production levels, proven coal reserves are estimated to last not more than 50 years. In the recoverable coal reserves, only about 4.5~7.0 percent can be mined by surface mining method under current mining technology. There are 13 surface mining coal areas located around northern China, Shanxi-Shannxi-western Inner-Mongolia, northeasten China, southwest and Xinjiang, and about 70% is lignite coal, the others are mostly steaming coal.

Surface mining of coal was started around 1900 in China and in 1949 (the founding year of the People's Republic of China) the total production of coal was only 32.43Mt, in the longer period of coal extraction from the beginning of the 20th century to 1980s, surface coal mining production was no more than 4% of the whole country's production rate because the most coal beds are not suitable for surface mining methods to exploit. By the end of the year 2010, the annual coal production rate reached 3.24 billion tons, of which surface mined production was about 9 percent.

2.2 Steeply pitching seams

Surface coal mining methods used in China from 1949 to 1980 were mostly open pit method to excavate steeply buried pitching seams, in which the shovel-train system was prevalent, the overburden was transported from the pits to external spoil dumping areas.

2.3 Flat buried seams

After 1980, in the new developed surface mines shovel truck system has been broadly selected for the overburden removal and shovel-truck with crushing plant and belt conveyor system for the coal transportation from pit to the washing plant or storage areas. Since the coal seams in those mines are usually flat or shallow buried, the first cut overburden of those mines is hauled out of the pits and then the following cuts overburden is dumped in the previous cuts.

3. Coal surface mining methods

The major types of surface mining methods[3,4] presently employed in China can be classified into: (1) Openpit mining; (2) Area Type Mining; and (3) Contour Mining.

3.1 Openpit mining[5,6]

Open pit mining method was used broadly in Chinese surface coal mines in the years from 1949 to 1980, these mines were characterized by dipping seams with the slope of more than 12 degrees to 45 degrees.

Open Pit Mining was almost exclusively employed in China's state owned surface mines with pitching or steeply pitching seams in flat, rolling or hilly topography. Open Pit Mining of coal seams (usually from 10 up to 100 meter thick) involved removal of overburden in flat terrain, rolling or hilly topography. Shovels of 3 to 4 cubic meter dippers were the most common equipment used to remove overburden and coal into a series of benches 8 to 10 meters high, and the haulage equipment was usually steam engine or DC driven locomotives with 7 to 14 railcars depending on the tractive effort of the engines. The ratio of overburden to coal varied from 1.0 to 10 cubic meters per ton of raw coal.

This type of Open Pit Mining is performed by having the stripping operation follow the coal down-dip, overburden is placed outside of the pit. Direct backfilling is not feasible in these kinds of mines.

Case study of openpit mining

Haizhou openpit coal mine is located at Liaoning province in northeast of China, started construction in 1950 and put into operation in 1953 with designed production rate of 3.0Mt/a to 5.0Mt/a in 1987, 3 Jurassic steaming coal seams mined of composition thick about 40 to 120 meters with an average 82 meters, with a dipping angle of 18 to 22 degree. The pit length is 4km from east to west and 2km from north to south and 350m in depth. Figure 1 shows the mine is under operation.

The mining system used in Haizhou mine was shovel train system with a dipper size from 4 to 10 cubic meters and the electrified locomotive with 60t side dumping railcars, the overburden was transported from pit to external dumping areas. Figure 2 illustrates the 150 ton DC locomotive used in the mine and figure 3 shows the power shovel with a dipper size of 4 cubic meters.

3.2 Area type mining

Area type mining in China is characterized by removal of one or two moderately thick to thicker coal seams (usually between 6 and 30 meters thick with some more than 100 meters),

Fig. 1. Haizhou Openpit coal mine.

Fig. 2. Electrified train locomotive used in Haizhou Openpit mine.

Fig. 3. Power shovel used in Haizhou Openpit coal mine.

with extraction activities ultimately progressing over a relatively large tract of land. The topography is usually flat or gently sloping, after the first cut overburden being dumped outside the pit, spoil is directly placed in adjacent previous cuts, using shovels and fleet of trucks for excavation.

In area type mining, coal is commonly hauled by fleet of trucks from the pit to crushing plant located outside the pit limit or directly loaded into mobile crusher on the working benches and then conveyed by belt conveyor to the washing plant or railroad station for out of mine transportation. In some mines, bucket wheel excavators and conveyors are used to extract soft loess strata and there is also a 90 cubic meter bucket dragline directly overcast the overburden into the adjacent previously mined out cuts with some pre-stripping of BWE and shovel truck combination. The overburden removal equipment utilized in these kinds of mines varies from dipper sizes from 10 to 60 cubic meters, the truck fleet payload from 32 to 362 metric tons. The coal loading equipment of single bucket machines varies from 10 to 55 cubic meters which is either electric or diesel engines.

Area type mining is broadly used in the surface coal mines established after 1980 with the coal seam of flat or gently sloping with varied topography, about more than 20 surface coal mines of over 10.00Mt/a designed production rate in operation or under construction.

Case study: An Tai Bao surface coal mine

An Tai Bao surface coal mine is located in Shuozhou, Shanxi province, started operation in 1985, with designed production rate of 15.00Mt/a dry raw coal. The average mineable thickness of the coal seams is 7 meters for No. 4 seam, 14 meters for the No.9 seam, and 3 meters for the No. 11 seam. The cover over the No. 4 seam varies from 80 to 120 meters thick. The interburden between No.4 and No. 9 seam is 35 to 45 meters thick, and the parting between the No.9 and the No. 11 seam is 6 to 10 meters thick. The average seam dip as a result of an anticline trending northeast and plunging southwest is from 2 to 6 degree. Figure 4 shows the total reserve area and the mining pits with the direction of mining indicated.

Excavation of pit 1 occurs in two phases, development and full production. The development phase is the mining work done prior to the preparation plant beginning full production. The full production phase is the time required to complete Pit 1 and to establish a normal haul back operation mining north in Pit 2.

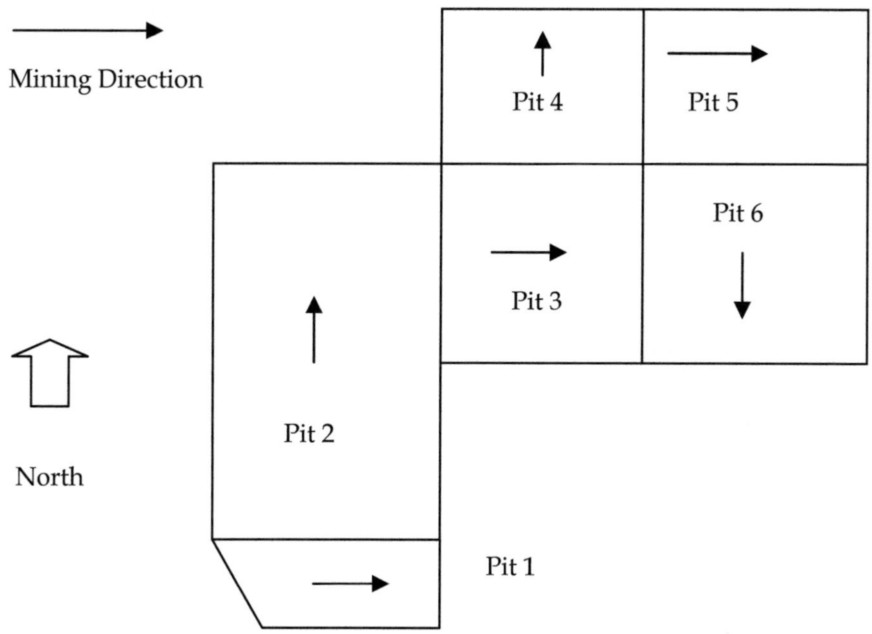

Fig. 4. An Tai Bao surface coal mine mining areas and pits.

During the development phase, all the excavated material must be hauled out of the pit to the waste area or the coal stock pile area. The full production phase of Pit 1 requires the waste material above the No.4 seam to be hauled to the waste areas outside of the pit and the waste bellow the No.4 seam to be backfilled in the mined out area of Pit 1. The coal is hauled to the raw coal dump at the plant site by the 154-tonne end dump trucks.

The excavation equipment includes the bucket of 25 cubic meters in size to be used in overburden removal, large hydraulic shovels, front end loaders in various sizes selected for coal loading and miscellaneous work.

3.3 Contour mining

Contour mining has been used only in Dafeng surface coal mine which located in Ningxia Hui Autonomous region northwest of China with mountainous topography. The working face of the mine is digging along the contour line to form the first cut and the spoil is hauled along the bench to fill in the nearby valley or hollows. Fig 6 shows the bench level of 1120m cut digging and waste materials hauled to fill nearby valley.

Fig. 5. An Tai Bao surface coal mine in operation.

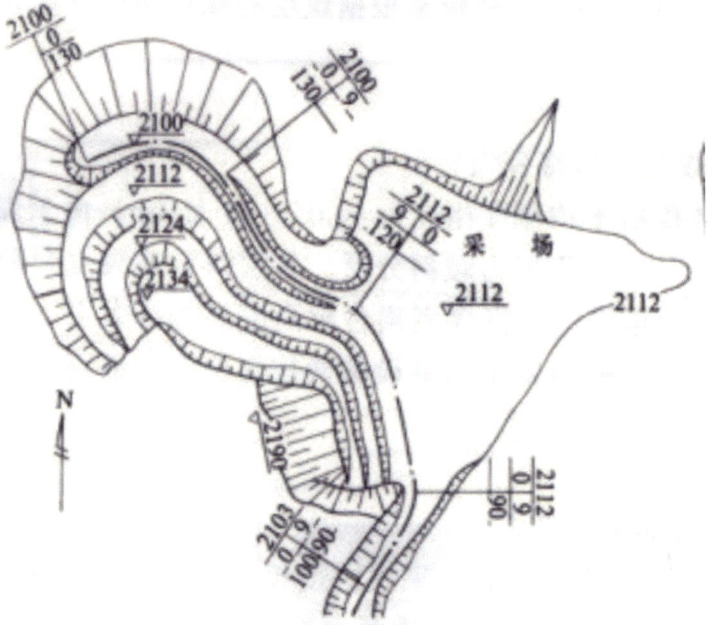

Fig. 6. From higher contour dropping to 2100m cut in Dafeng mine.

Dafeng mine is designed with annual output 0.9Mt anthracite and power shovel of 3 to 4 cubic meter dipper size for loading on 27 to 32, 45 ton off-highway trucks. The coal seams total thickness is 38.77 meters with a dipping angle from 5 to 17 degrees.

"Mountaintop removal mining" (MTR) is also used in this mine that uses explosives to blast "overburden" off the top of some higher mountains.

Fig 7 shows the large scale throw blasting used to move mountain top for underneath coal extraction.

Fig. 7. Throw blasting in Dafeng mine.

4. References

[1] National Bureau Statistics of China, Statistic yearly data, 2011.2.
[2] Weitong, Zhangxianchen, Wangyujun et al, Research of the strategy of China's coal development, Shanxi science and technology press, 1995.11.
[3] Kennedy B A. Surface Mining, 2nd Edition, Society of Mining, Metallurgy and Exploration Inc., Littleton, Colorado. 1990.

[4] Fung R, Surface Coal Mining Technology, Engineering and Environment Aspects, Noyes Data Corporation. Park Ridge. New Jersey. USA. 1981.

[5] Atkinson. T, Burdon R. G., Advanced Surface Mining Technology Project Report, The University of Sydney, 1984.

[6] Martin C H, Australasian Coal Mining Practice, The Australasian Institute of Mining & Metallurgy. 1986.

Part 2

Underground Mining

3

Ground Control for Underground Evaporite Mine in Turkey

T. Onargan[1], K. Kucuk[1], A. Deliormanli[1], S. Saydam[2] and M.Y. Koca[3]
[1]*Faculty of Engineering, Department of Mining Engineering, Dokuz Eylul University, Buca, Izmir,*
[2]*University of New South Wales,*
[3]*Department of Geological Engineering, Dokuz Eylul University, Buca, Izmir,*
[1,3]*Turkey*
[2]*Australia*

1. Introduction

Ground control is the methodology applied to maintain all the risks associated with various forms of ground movement and inundation in underground mines within an acceptable level. It is applied to all stages of a mine - from feasibility through operation and finally abandonment.

Ground control methodology is largely determined as a function of the interaction of various qualities of the rock mass with various aspects of the mine planning and design methodologies. Depending on the nature of these interactions, rock support and reinforcement will required to achieve effective ground control.

Consequently, effective ground control may be considered to be a function of three main components:

- Site ground characteristics,
- Mine planning and design; and
- Ground support and reinforcement

Controlling the potential for hazardous ground movements and inundation of an underground mine to within acceptable limits is essential. Both hazards can result in serious harm or death of workers or persons that may inadvertently enter a mining area. The hazards are not always obvious. For example, the outcome of the hazard of a loose rock falling from a sidewall and striking someone can be fatal by either direct physical contact, or damaging the plant in which the worker is working. The presence of certain geological structure (natural planes of weakness in the rock) that override the effectiveness of any ground control measure, is not always obvious from within the mine.

In the case study given in this chapter, in situ and laboratory studies have been made to define the formations through which the drift has been driven. In situ engineering geology measurements were carried out consisting of field observation, mapping, boreholes and laboratory tests on samples collected from the trona field and the boreholes. After drift excavation and support installation, the deformations and the loads on the supports were measured. As a result of this study, the deformational characteristics of the trona ore beds and weak rocks and their effect on the main drift (T-2000) deformation have been determined. In this study, some parameters for understanding the deformational behavior of trona are investigated including in situ rock support loads, rock properties, and geology and excavation sequence.

2. Background for failure mechanism of underground evaporite mines

Geology is one of the most important parameters determining the overall deformational behavior of an underground opening. With careful investigation of the geological details, it is often possible to explain the unexpected deformational behavior of an excavation(Wang et all, 2000). Deformation of underground salt, trona and potash mines is generally time dependent, providing for gradual adjustment of strata to mining induced stresses. Time dependence can allow for higher extraction ratios provided eventual failure can be tolerated. However, this eventual failure can be violent if creep deformation can shift stress and potential energy to strong, brittle geologic units.

Generally, room and pillar methods, short-wall methods and solution mining methods have been applied as production methods for all evaporite mines in the world. The mine failure case studies reviewed here illustrate this process. Yield pillars and defects in bridging strata figure prominently in these cases. Yield pillars provide local and temporary support to the roof, temporarily delaying the cave; and allowing extraction ratios and overburden spans to increase beyond the long term capacity of overlying strata. Defects (faults, voids, thinning) of strong overburden strata reduce the critical span, sometimes to less than panel width. Analyses of many of these cases have focused on a cascading pillar failure mechanism, but recent work and this review point to failure of strong overburden strata as the essential element. The suddenness of failure and attendant seismic events pose hazards to miners and, in some cases, to those on the surface(Whyatt, J., 1998).

Trona is generally much stronger and stiffer than the immediate roof and floor rocks, with the degree of contrast varying with different beds and locations. Trona shows some time-dependent behavior, but few laboratory creep tests have been performed in this study. Much of the time-dependency exhibited in the field may in fact represent creep of weaker roof and the floor strata rather than the trona bed itself. Many researchers suggest that, if the deformations around underground excavations are carefully measured in situ, the field measurements provide a better understanding of the deformational behavior of underground excavations(Onargan et al, 2006).

3. Case study: Beypazari trona mine

3.1 Location and geology

The study area is near Beypazari town that is 100km northwest of Ankara, and the Beypazari natural soda (trona) field is located in an area of 8 km^2, which is 20 km northwest of Beypazari town center (Fig. 1)

Beypazari natural soda (trona) field exists in the lower parts of the volcano-sedimentary sequence of the neogene basin and characteristically comprises no outcrops. The elevations

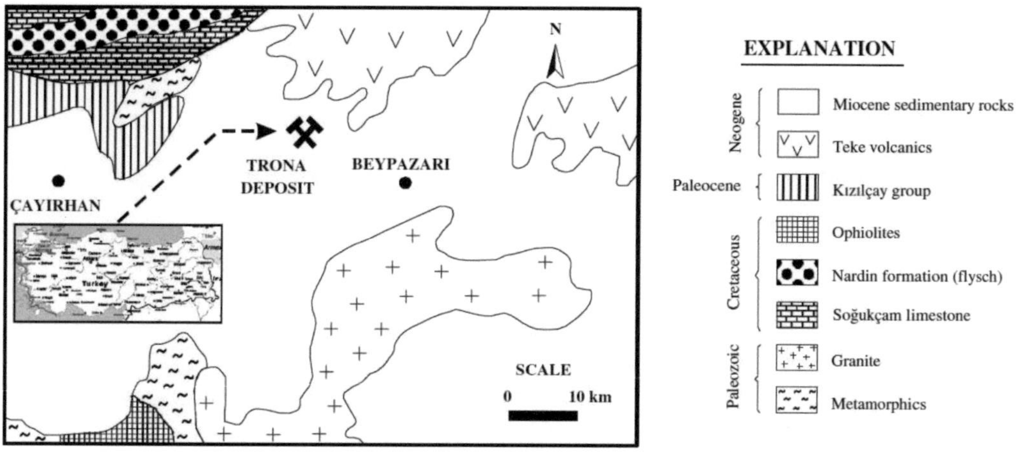

Fig. 1. General geology and location map of the study area (from Helvaci; Helvacı et all, 1989; Aksoy et all,2006).

in the Beypazari Neogene basin change between 800 and 1100 m and there is often a smooth morphology in this area. Continental climate dominates the region and there is no active stream system present there. The Neogene rock units are dominantly precipitated in shallow pond domain. The probable source of the Na ion required for the accumulation of trona deposit is the extended Neogene igneous rocks compacted with the sediments in the northeast of the basin. The Beypazari natural soda deposit is structurally confined by two fault systems and split by another fault system. Of those faults, the Zaviye fault has developed as a consequence of an extended tectonic regime affecting the region during the Lower-Mid Miocene. The Cakiloba fault is a folded series of an anticline- synclinal sequence. Its general orientation is N73E. This folding can also be described as a highly inclined monocline extending along NW-SE (Unal et all, 1997). The Kanliceviz fault splitting the trona deposit is generally oriented N20W and it has a dip-strike of 35-600 SW. It intersects the Zaviye and Cakiloba faults almost perpendicularly and has the character of an inverse fault(Figure 2.).

3.2 Laboratory studies

This study is divided into two sections: the first one includes identification of clay samples; the second one covers determination of the geotechnical parameters of the soft rocks and soda beds around the main drift. Rock types which occurred along the main drift line, based on thin sections and X-ray diffraction analyses, which were carried out in accordance with the methods described by the ISRM. Different block samples obtained from the tunnel face have been used in this study. The test results are summarized in Table 1.

The study area consists of two different soda zones with intermediate bands in between. The upper zone includes six separated soda beds. The dominant soda mineral in U1 and U2 is nahcolite; U3 is combined of trona and nahcolite; U4-U5 and U6 consist of mainly trona minerals. Since the natural soda bed U2 is too weak and thin, reliable results could not be obtained. While limited numbers of tests were carried out for the U1 and U3 beds, detailed tests have been conducted for the U4, U5 and U6 beds.

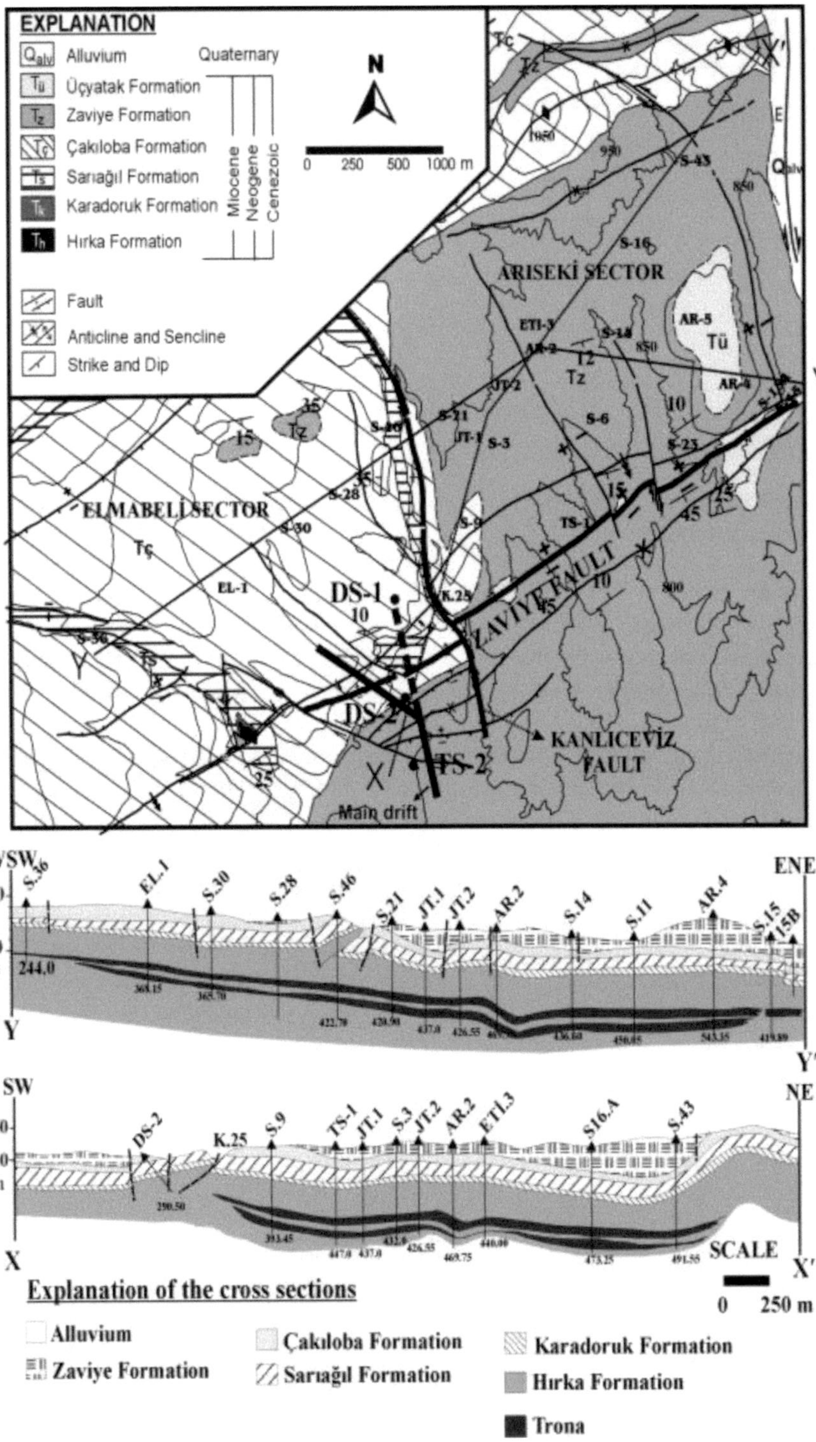

Fig. 2. Geological map and cross-sections of the Beypazari natural soda field (Onargan et all, 2004).

Engineering properties	Unit	Hanging walls	Soda beds
Unit volume weight	Kg/m^3	1960–2500	1820–2410
Porosity	%	1.61–11.71	0.47–2.35
Water content, ω	%	6.69–21.22	—
Slake durability, I_{d2}	%	11.24–69.56	6.7–32.60
Point load index, I_p	MPa	0.29–1.33	0.46–1.81
Uniaxial compressive strength, σ_c	MPa	6.9–32.02	5.0–46.2
Young Modulus, E	MPa	900–14800	900–26100
Cohesion, c	MPa	0.29–3.3	0.90–8.60
Internal friction angle, ϕ	(°)	42–65	48–63

Table 1. Geomechanical properties of hanging walls and soda beds (Onargan et all, 2004).

The geomechanical properties show great variety in the soda-bearing-bituminous shale formations due to the soda content. Two graphs illustrating the mechanical strength and deformability values obtained for the waste rocks and soda beds are given in Fig. 3. As seen in these diagrams, both the soda beds and waste rocks show a broad range of deformation properties, varying from "very weak" to "medium" characteristics.

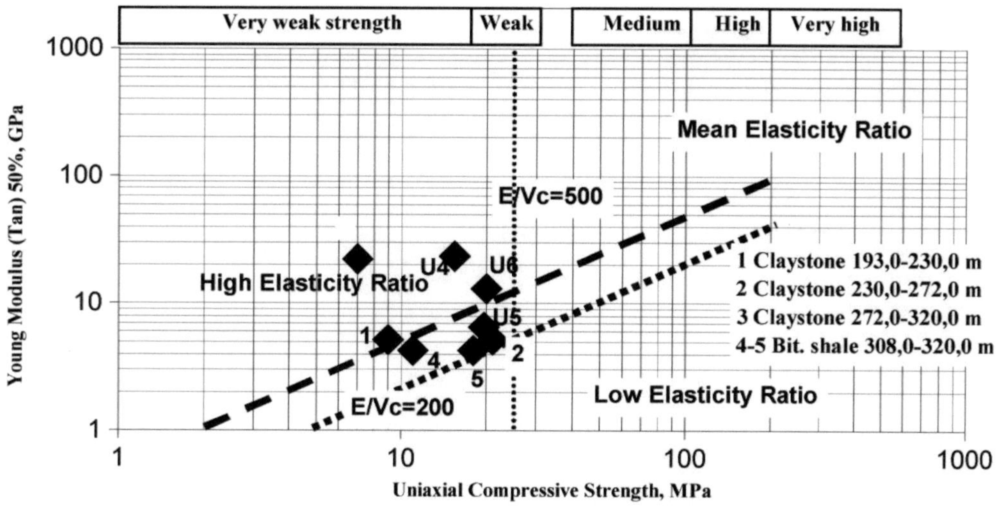

Fig. 3. The relation between uniaxial compressive strength and Young's modulus for the soda beds and waste rocks (Onargan et all, 2004; Onargan et all, 2001).

3.2.1 Creep test results

The time-dependent deformation of trona, or creep, consists of four stages, namely: an elastic deformation stage, a transient or primary creep stage, a steady state or secondary creep stage, and a tertiary creep stage. The instantaneous elastic deformation occurs immediately after the creation of an excavation and is followed by the primary creep stage.

In this primary creep stage, a high deformation rate decreases exponentially with time. The exponential decrease of the deformation rate under constant load in laboratory creep tests can be described as a strain hardening effect.

In this study, the creep tests have been carried out to determine the deformability properties of rocks versus time. The samples were tested under a load equal to 75% of the uniaxial compressive strength at 20 °C room temperature. The deformability properties of clay Stone and trona beds (U4, U5, U6) versus elapsed time have been recorded. The creep properties of clay stone, U4 and U5 seem weaker when compared to U6. The clay stone, U4 and U5 ores complete the behavior of the 3rd creep region in 800–1000 min, while the U6 ore completes the same region in about 1400 min. (Figure 4.)

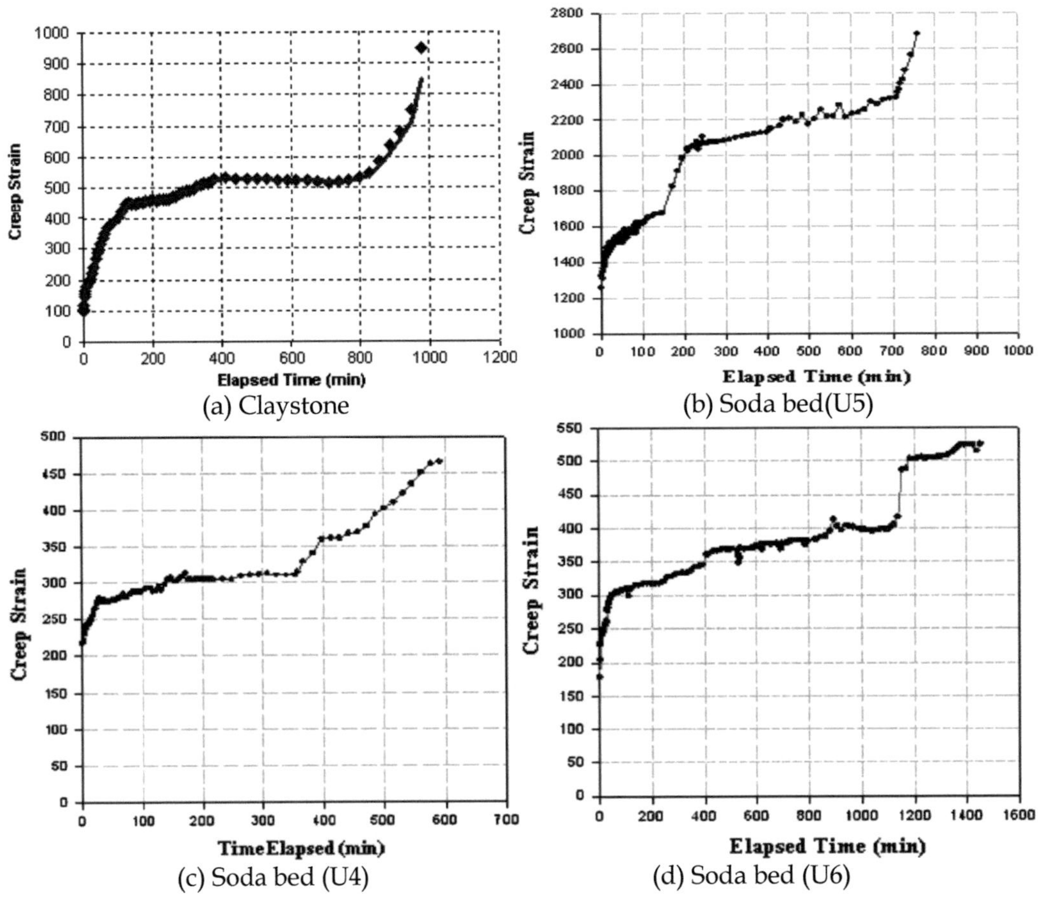

Fig. 4. Creep versus time for rock and soda samples(Onargan et all, 2004).

3.2.2 Atterberg limits and swelling tests

The dominating mineral within the clayey formations is montmorillonite accompanied by a trace amount of illite. The swelling properties of the intermediate clay bands and consolidated clay have been found as 4–18%. For the intermediate roof formation of U1, the swelling coefficient under pressure was found to be 7% as tested by the oedometer. This clay formation is termed Lithology 2 in this study and the intermediate clayey band between U1 and U2 is termed Lithology 5. They both show the characteristics of CH type of clay. Therefore, these levels require special attention during the mine planning. Apart from these, the clay Stone between U2 and U3 are termed Lithology 8, Lithologies 11 and 12 between U3 and U4, and Lithology 21 taking part between U4 and U5, are all in the category of clays of high plasticity. Several tests, have been conducted for the predetermination of index values on the intermediate bands and hanging wall formations. These formations are made up of clay stones, and they show soil character due to the weathering and existence of water.

Atterberg limits of these clayey soils were determined by a procedure suggested by ASTM. The Liquid Limit values (LL) vary between 38 and 88 for clay and soft clay stones, while the Plastic Limit (PL) values are between 27.81 and 56.25 and the Plasticity Indices (PI) were found to be about 10–40. According to the Unified Soil Classification Systems (USC); these clays, are classified as CH, and they are also under the group termed "Inorganic Oily Clays with High Plasticity". In the tests applied to un-altered samples, the consolidation index and swelling index were determined as 20% and 7%, respectively. These kinds of soils are classified as "high swelling potential soils" by O'Neil and Poormoayed. Hence, it is foreseen that additional support precautions will have to be taken at the necessary positions in the underground openings.

3.3 Field measurements at the T-2000 main drift and T-3000 roadway

The Main Drift, T-2000, and a roadway, T-3000, were excavated for Beypazari trona (natural soda) underground mine project between the years 1999 and 2001. The instruments were installed at several underground monitoring stations in order to obtain the necessary in situ data regarding displacement, deformation of supports and load using the load cell system. The magnitude and orientation of the loads were analyzed according to the results of the in situ measurements. Rock pressure was determined by the in situ measurements of the deformation of the support.

Support load and convergence measurements were recorded daily for a period of about one year until the convergence rate decreased significantly. Six different stations were set up along the main entry drift for these measurements.

Load cells were placed at some of the points and the load increase over the support units was determined versus time. Three more stations (U4-a, U4-b and U4-c) in the roadway (T-3000) have been established within the soda bed (U4), again to determine the amount of load and convergence.

The stations in the main drift were coded as K-20, K-21, K-22, K-23, K-24 and K-25, which were all after the support installation. Vertical and lateral closures and narrowing were monitored daily for a period of 400 days.

3.3.1 Support loads

The support system load values were recorded by a digital gauge connected to the load cell system. The measured support loads at the stations versus time are plotted and shown in Fig. 6. According to these measurements, the support load at K-23 roof-rock station was found to be 0.09 MPa. At the U4-a station in the soda bed, the support load was 0.06 MPa.

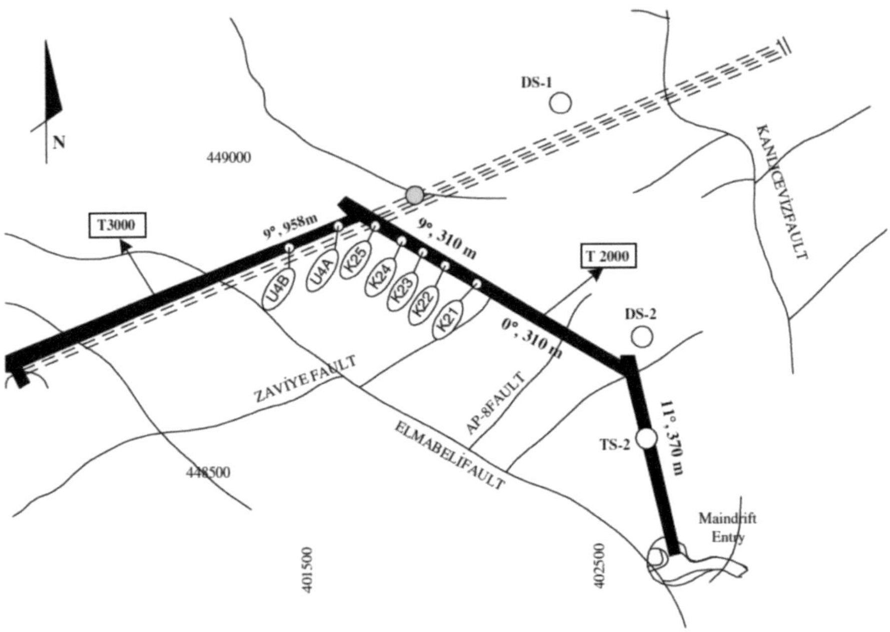

Fig. 5. Monitoring station and plan view of the main drift and underground openings in Trona mine (Onargan et all, 2004; Aksoy et all,2006; Onargan et all, 2001).

These values are quite close, because the strengths of the formations are similar and the spacings of the steel sets are also closely selected. For some locations, the support loads were measured as 0.31 and 0.24 MPa, respectively for the steel set intervals of 1 and 0.70 m.

3.3.2 Convergence monitoring

Convergence measurements at the main drift revealed variations in measured values over long periods of time. These values are independent of the face advance. Nine measuring stations were considered suitable for the analysis of the measured data. In these stations, horizontal and vertical convergences were measured between points of the rock wall.

This measurement system is illustrated in Figure 7 and 8. Between the tunnel length of 679 and 813 m, the main drift was driven in clay stone, bituminous shale, volcanic tuff and tuffitic claystone. In this part, 'rigid arches' were used as the supporting system by the management. For this section of the drift in the clayey formation, two kinds of problems were faced. Firstly, large deformations took place. Secondly, considerable swelling was observed, which was expected in the silty clay formations as mentioned before.

Floor heaving due to swelling capacity of the weak formation was also observed. The convergence measurements were performed using a tube-extensometer. The details of these measurements can be found in the research report(Onargan et all, 2001). All the resulting values indicate that deformations were below 1% on the convergence measurement lines Similar tests, for all the stations chosen, were carried out in different formations. Results can be seen in Figs. 9 and 10 in graphical form. It can be deduced from the foregoing studies that the total convergences at all the stations were found to be below 1% of the total gallery height and deformation rates were reduced significantly.

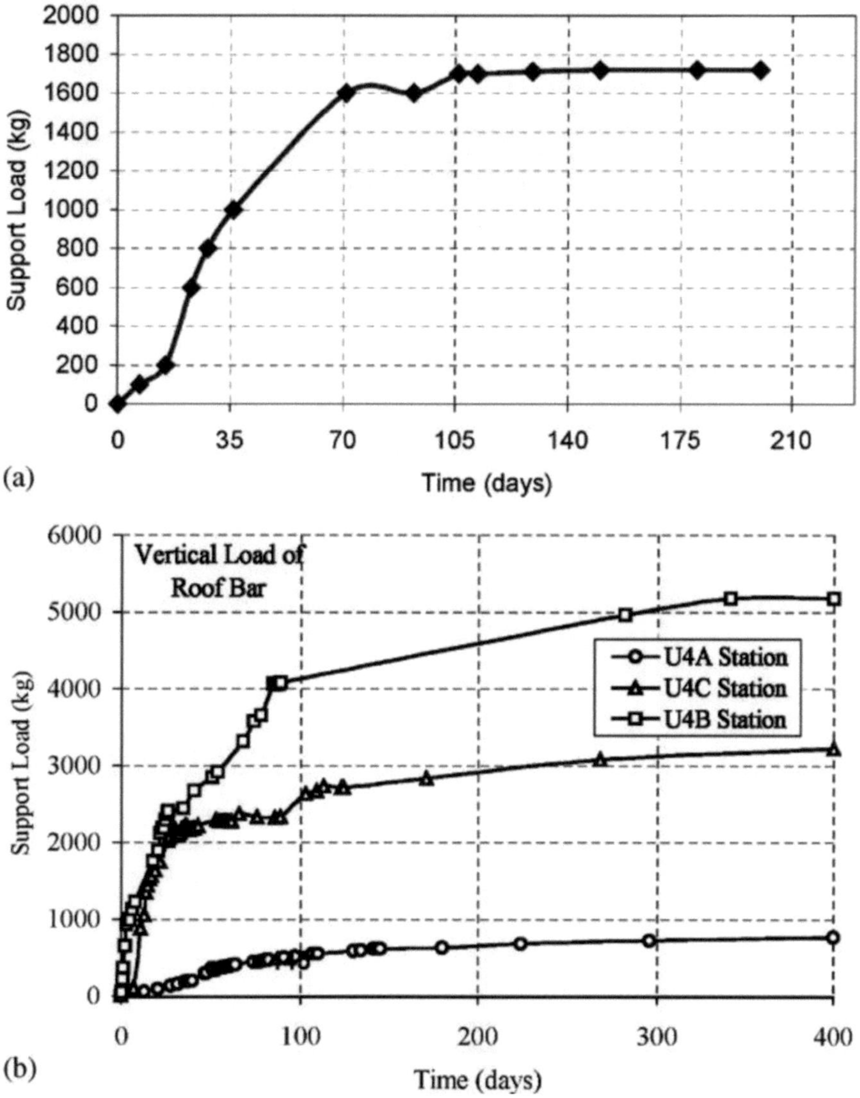

Fig. 6. Support loads versus time at soda beds and roof rock stations: (a) roof rock station (Station number: K-23) and (b) soda beds stations. (Onargan et all, 2004).

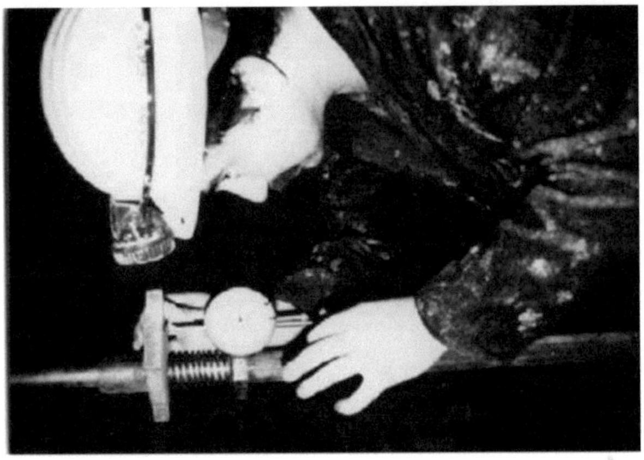

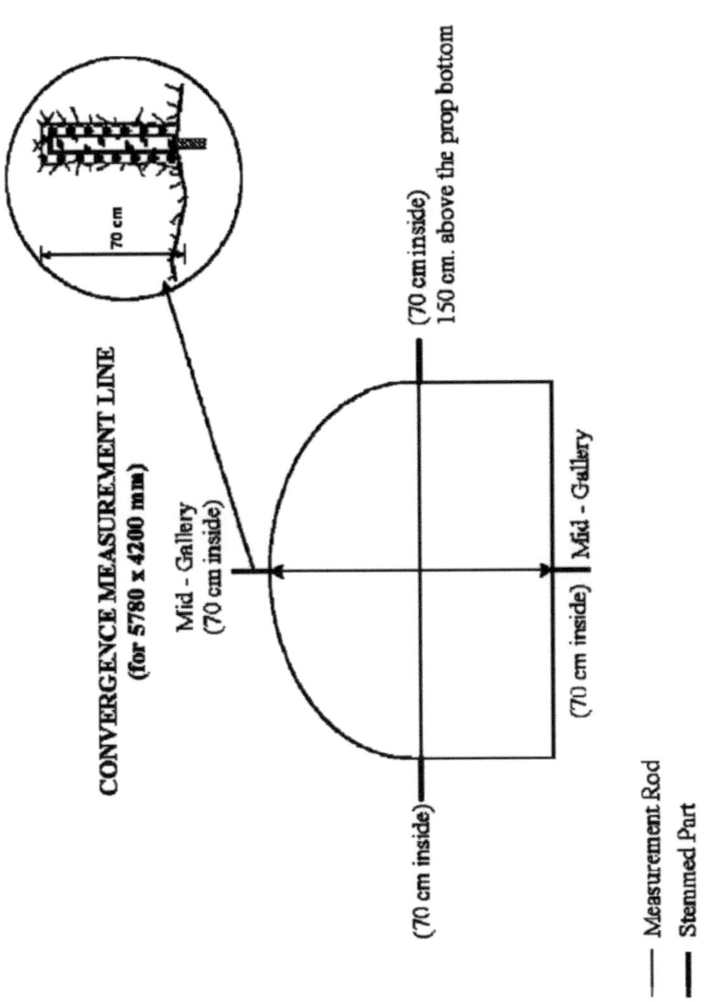

Fig. 7. Lines and intervals of convergence measurement at the stations(Onargan et all, 2004).

(a)

(b)

Fig. 8. Some views of the measurement stations in the hanging-walls: (a) a view of the station (K-24) and (b) convergence and support load measurement studies(Onargan et all, 2004).

In order to study the behavior of the support system, the convergence measuring stations were installed in the roadway (T-3000) driven in the soda bed. This roadway was driven with a 18 m2 cross section, using yielding steel sets of 29 kg/m TH profile with several spacings. These stations are termed U4-a, U4-b, and U4-c and convergence measurement and support loads were measured daily over a period of 400 days. As can be seen from Fig. 10, the maximum vertical closure was recorded as 198 mm, 5% of the original roadway height at the U4-a station. The values were obtained as 340mm (8.5%) at U4-b, and 198mm (5%) at U4-c for the other monitoring stations.

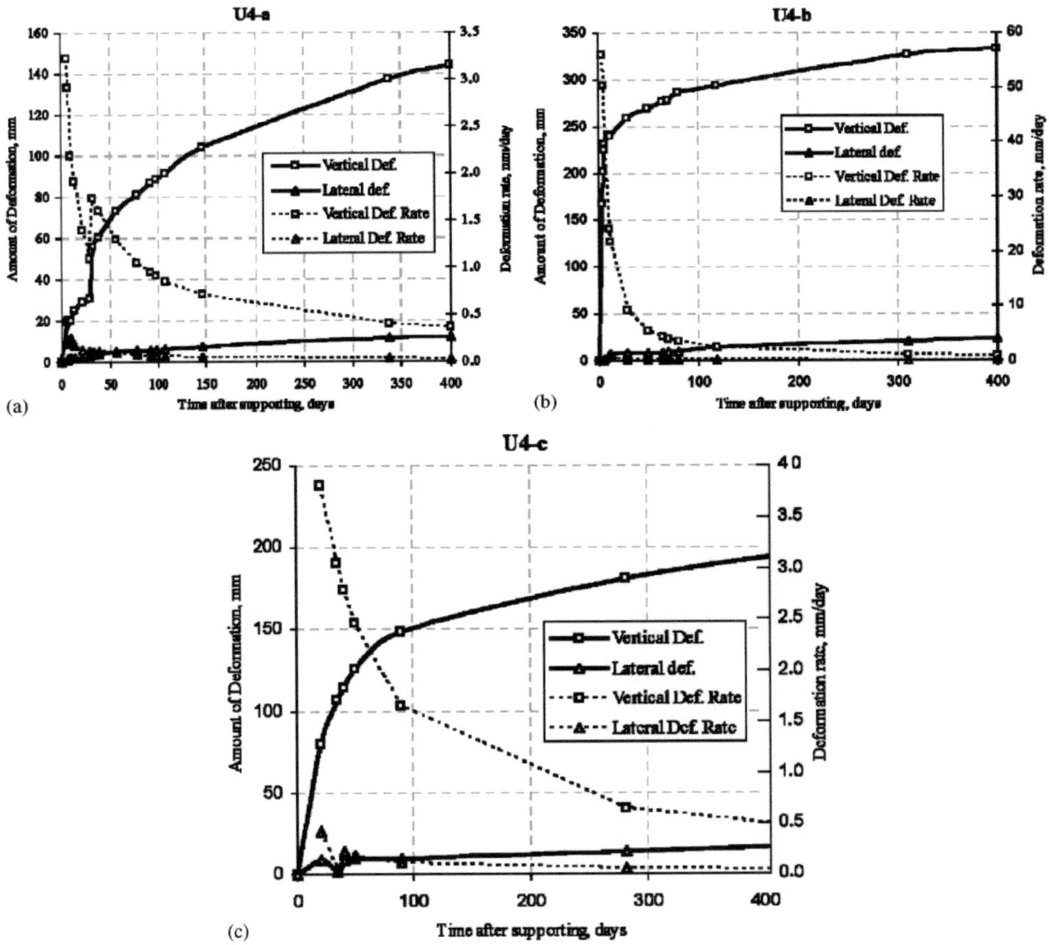

Fig. 9. Convergence versus time at the soda beds: (a) U4-a station (Steel set spacing 0.5 m), (b) U4-b station (Steel set spacing 1.00 m) and (c) U4-c station (steel set spacing 0.70 m). (Onargan et all, 2004).

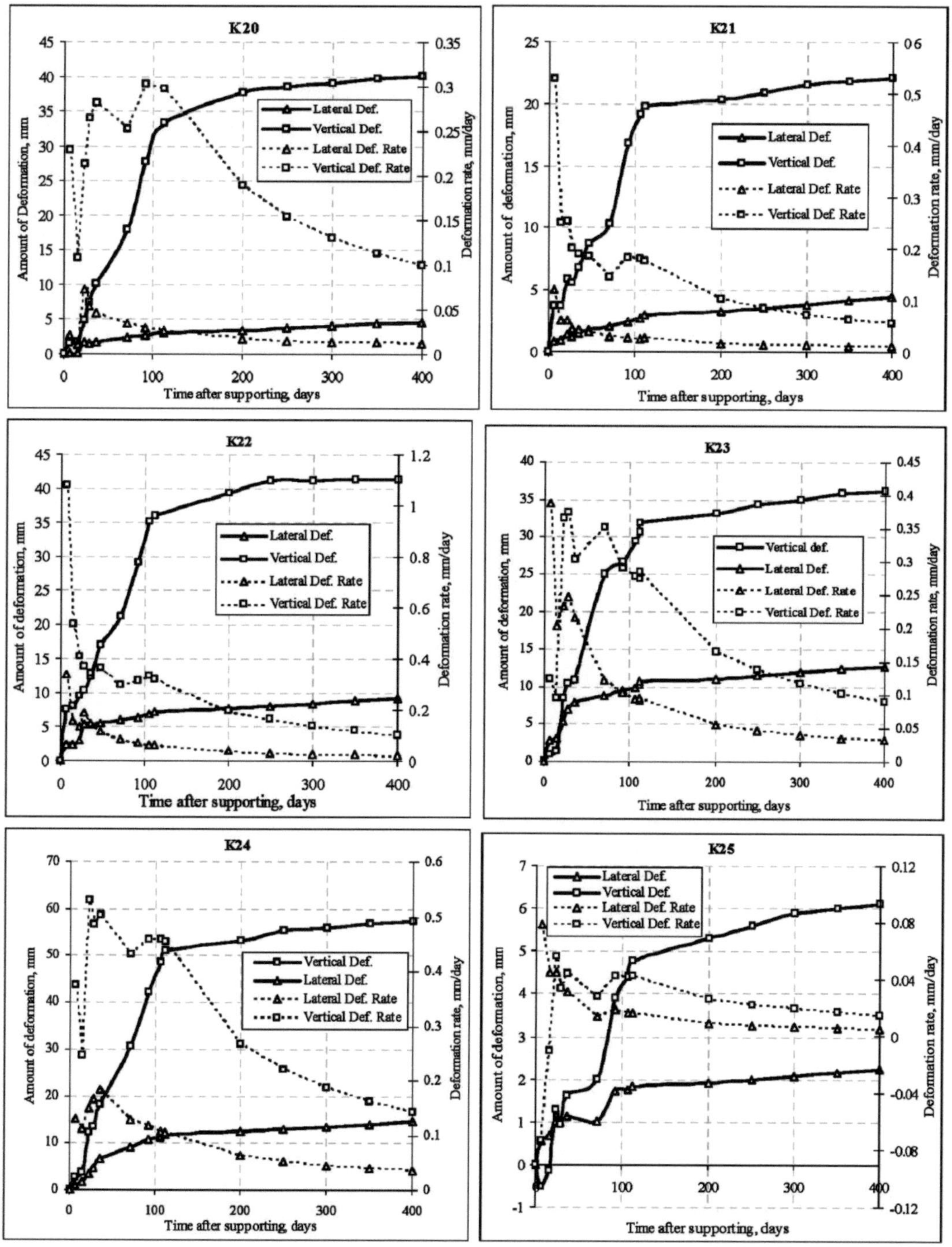

Fig. 10. Convergence versus time at the hanging wall stations (Onargan et all, 2004).

4. References

[1] Aksoy CO; Onargan T; Yenice H; et al. Determining the stress and convergence at Beypazari trona field by three-dimensional elastic-plastic finite element analysis: A case study, International Journal Of Rock Mechanics And Mining Sciences, Volume: 43 Issue: 2 Pages: 166-178 DOI: 10.1016/j.ijrmms.2005.06.004 Published: FEB 2006.

[2] Helvaci C, Inci U, Yilmaz H, Yagmurlu F. Geology and Neogene trona deposit of the Beypazari region, Turkey. Turk J Eng Environ Sci 1989;13(2):245–56 (Turkey).

[3] Helvaci C. The Beypazari trona deposit, Ankara Province, Turkey. In: Dyni JR, Jones RW, editors. Proceedings of the first international soda ash conference, Vol. II. Larami: Wyoming; 1998. p. 67–104.

[4] Onargan T; Koca MY; Kucuk K; et al., Impact of the mechanical characteristics of weak rocks and trona ore beds on the main drift deformation at the beypazari mine, Turkey. International Journal Of Rock Mechanics And Mining Sciences, Volume: 41 Issue: 4, Pages: 641-654 DOI: 10.1016/j.ijrmms.2004.01.006, Published: JUN 2004

[5] Onargan T, Koca MY, Yenice H, Gungor T, Kokturk U, Deliormanli AH, Saydam S, Aksoy O, Kincal C. The Geotechnical study for the main drift of Beypazari–Ankara trona field. Research Report, D.E.U. Engineering Faculty, Mining Engineering Department and Project Number: 99-02, Izmir, Izmir-Turkey, 1999, 200p.

[6] Onargan T, Koca MY, Kokturk U, Yenice H, Gundogan I, Kahraman B, Deliormanli AH, Saydam S, Karakus D, Kucuk K, Aksoy O. ve Pamukcu C. Geotechnical support design for main haulage road at Beypazari trona mine, Turkey. Project number: DEU-MAG-20011, Izmir-Turkey, 2001, 156p.

[7] Unal E, Ulusay R, Ozkan I. Rock engineering evaluations and rock mass classification at Beypazari trona site. Middle East Technical University, Mining Department, Technical Report, December, Ankara, Turkey, 1997.

[8] Wang C, Wang Y, Lu S. Deformational behaviour of roadways in soft rocks in underground coal mines and principles for stability control. Int J Rock Mech Min Sci 2000;37:937–46 (Elsevier Science).

[9] Whyatt, J and Varley, F. Catastrophic Failures of Underground Evaporite Mines. NIOSH - Spokane Research Laboratory, USA.

4

Longhole Stoping at the Asikoy Underground Copper Mine in Turkey

Alper Gönen
Mining Engineering Department, Dokuz Eylul University, İzmir, Turkey

1. Introduction

Asikoy underground copper mine is located in Küre county, some 60 km north of Kastamonu province and 25 km from Black Sea cost. Underground mining method is longhole stoping with post backfill. Ore production is 420.000 ton/year with an average of %2 copper grade[1].

2. Geology of the region

Küre copper deposits occur along the middle pontide zone. Although it exists in a region which has considerably different geological past from the southeast Anatolian ophiolite zone, Küre massif sulfide deposits comprise properties that could be classified within the Kieslager type, which is between Cyprus type and Kuroko type [2]. In the region, there exist the plagical sediments formed of subgrovacs and shales and also the toleitic basalt volcanites that are the products of the mid-ocean extension. It is seen that important tectonic movements occurred within the Küre formation. The units are intercepted with an N-S oriented fault. The mineralization takes place in the weak zone induced by this fault and also within the toleitic basalts and along the borders of the plagical sediments. The overall geologic map of the study area is demonstrated in Figure 1.

The ore mass occurs within the altered basalt series that are a part of the Küre ophiolites and is overlain with black shale. The ore mass consists of coarse lenses broken with faults and thrusted. The ore, which is composed of pyrite and chalcopyrite, is in the form of massif lenses with high grades under the hanging wall black shale and in the form of stock work pyrite and chalcopyrite veins with low grades within the altered footwall formation. Pyrites and chalcopyrites occasionally show colloform textures. (Figure 1).

3. Underground mining method

In Aşıköy underground copper mine, there are two main mineralisation zones named as the east and west sector between the 945 m level and 792 m level. The main access adit is horizontal and connects with spiral ramp at 932 m level. Spiral ramp that developed in the

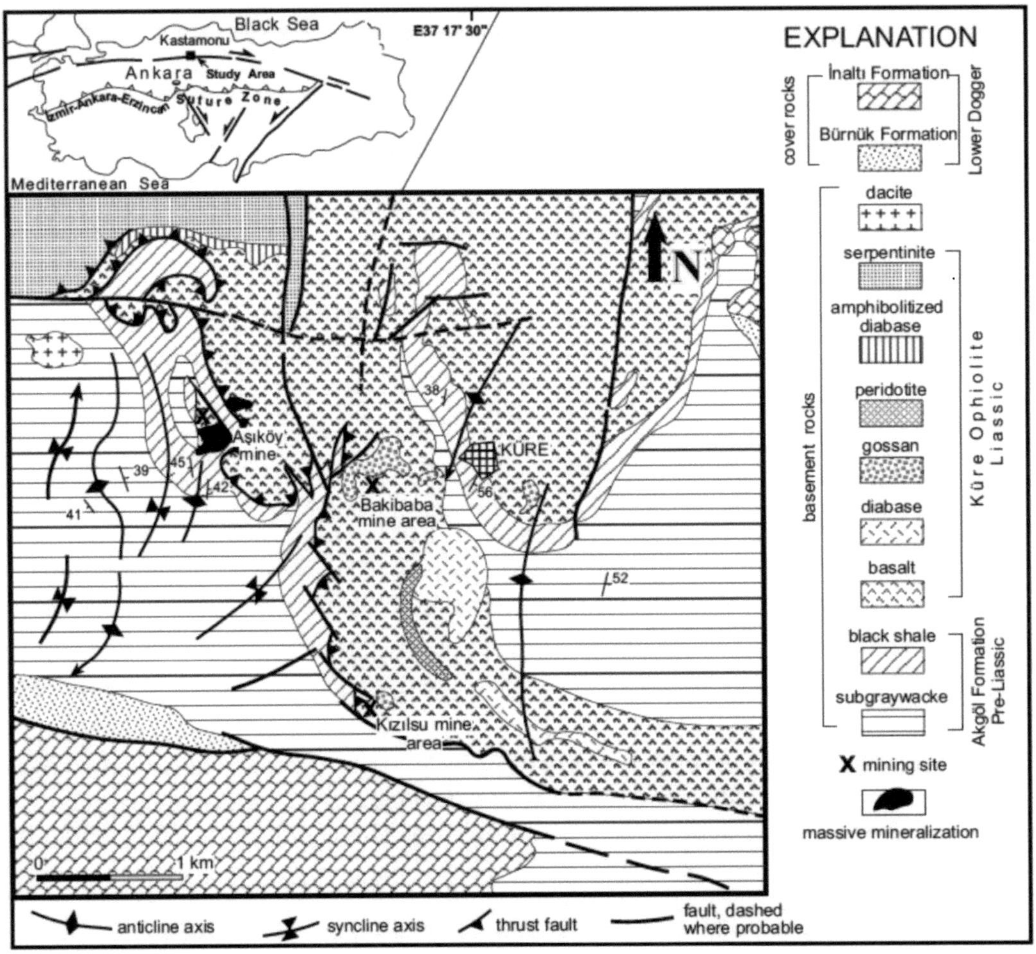

Fig. 1. Simplified geological map of the Küre mining district (Kuşçu and Erler, 2002) [3].

footwall of the orebody was driven between 932 m level and 792 m level at 6-7 degrees (Figure 2). There are two vertical shafts from surface (1080 m level) to bottom level for ventilation. The exhaust shaft is equipped with a winch-raised cover, which can be raised in winter to induce natural air movement.

Most development is within the competent footwall rock mass. Drilling rounds are about 53 holes, 45 mm in diameter and 3.5 m hole length. The orebody exhibits different rock mass characteristics. Ground support is by shotcrete, bolting with mesh, mesh reinforced shotcrete, standard Swellex in 2.4 m and 3.3 m lengths, and cement grouted bolts in 3 m, 4 m and 6 m lengths.

Orebody, which dips at 60 degrees, is accessed from this ramp, along the levels that are spaced 12 m in vertical interval. At each level, along footwall contact or in the center of the orebody, strike access drifts are developed with 4.5 m-wide x 4.5 m-high dimension. Across the strike, 7 m-wide x 4.5 m-high sill drifts are driven until the hangingwall contact. These drifts vary in length, depending on the thickness of the orebody [1].

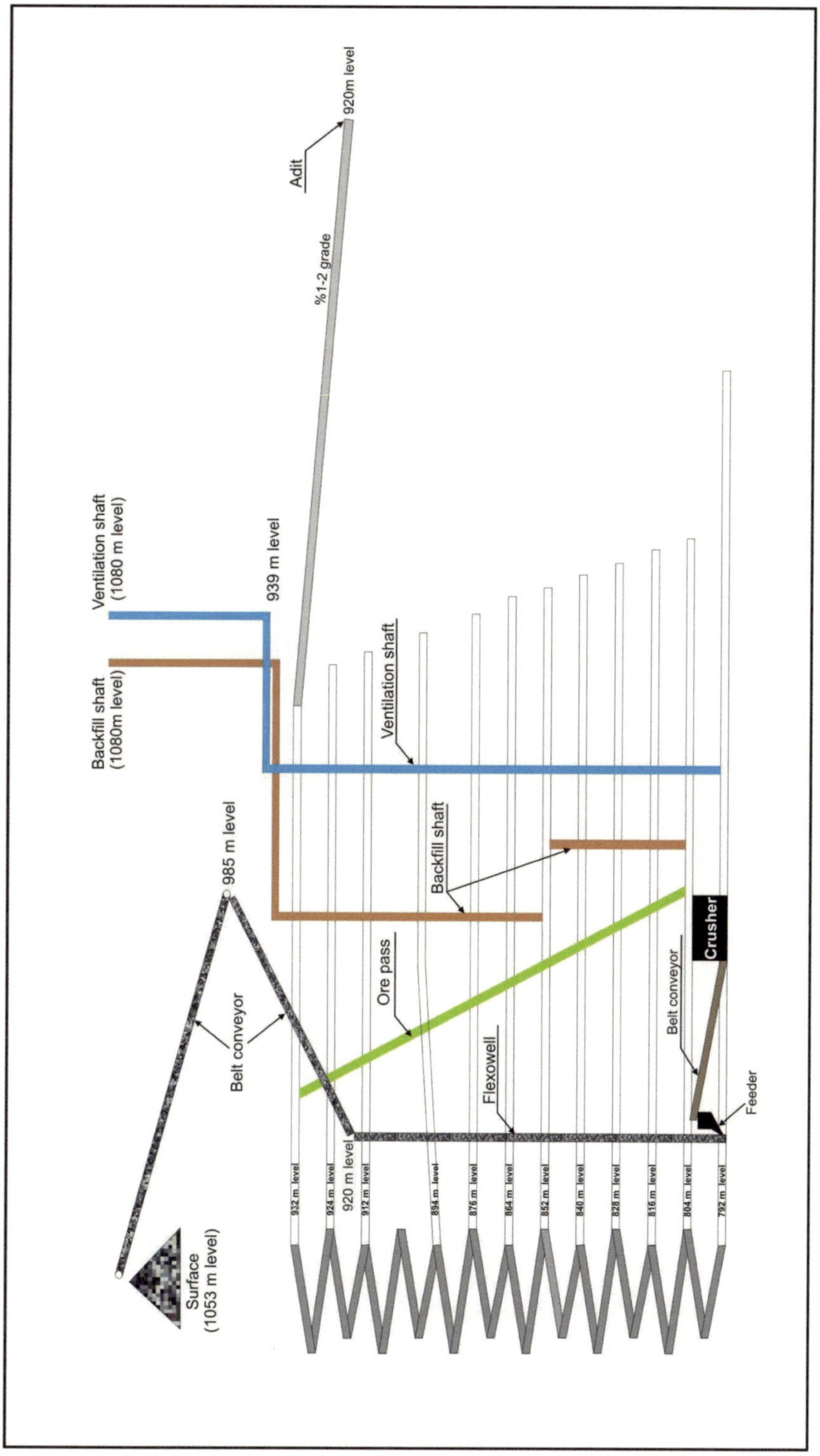

Fig. 2. General underground mine layout.

At the end of the sill drifts, 1.5 m x 1.5 m slot raise is opened and then widened out to 7 m to drift width. Blast holes are drilled parallel with 76 mm diameter in the downward direction between two sill drifts. Hole lengths are 7.5 m and one or two rows are blasted at a time. Spacing and burden between holes are 2.1 m and 1.9 m respectively. The main blasting agent is ANFO, with powergel primers and nonel initiation (Figure 3).

Blasted ore is mucked from lower sill drift by remote controlled LHD and transported to orepass. At 804 m level, orepass system feeds the underground crusher. -10 cm crushed ore travels along a conveyor belt to a feeder and into flexowell vertical conveyor belt system at 792 m level. At 920 m level, a belt conveyor at an avarage grade of 8 degrees transfers the ore to the surface primary crusher.

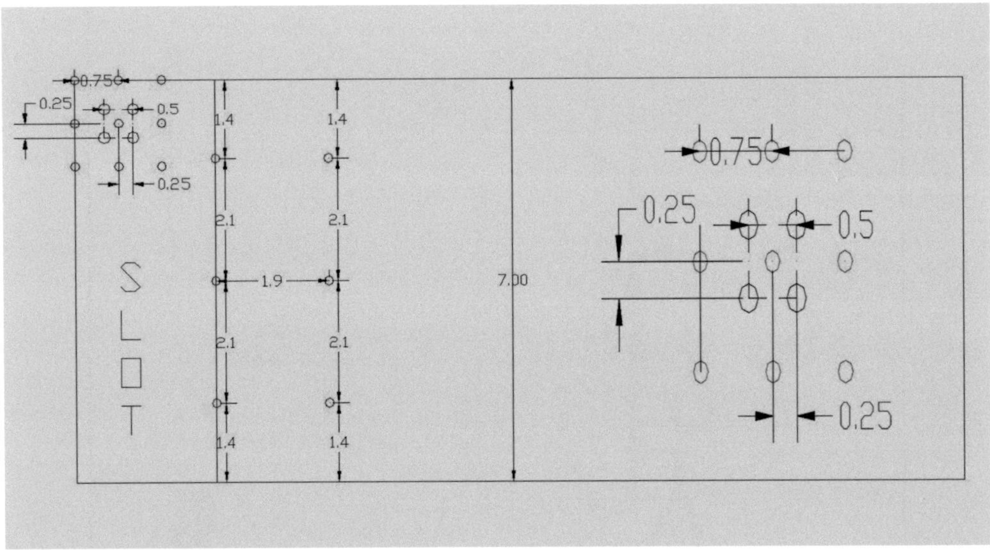

Fig. 3. Slot opening at the end of the stope.

After extracting the ore between two sill drifts, the open stope is backfilled from upper sill drift (Figure 4). Two types of backfill material are used. These are cemented rock fill and uncemented rock fill. Cemented rock fill has % 5 cement content by weight and is used for backfilling of primary stopes. After two adjacent primary stopes are backfilled, the primary pillar between them can be mined as secondary stope and backfilled with uncemented rock fill. Trucks are used for both types of backfilling. Open stopes, especially located at orebody boundaries, are backfilled with cemented rock fill.

4. Mine stability and mining sequence

In order to analyse the stability of the stopes and pillars and the the whole mine in general, a series of rock mechanics tests were performed on the core samples taken from the ore, basalt and backfill material and the geomechanical properties of the samples were determined as in Table 1. In the models set up by using Phase^2 program, namely the finite element method, these values obtained in the laboratory were used [4].

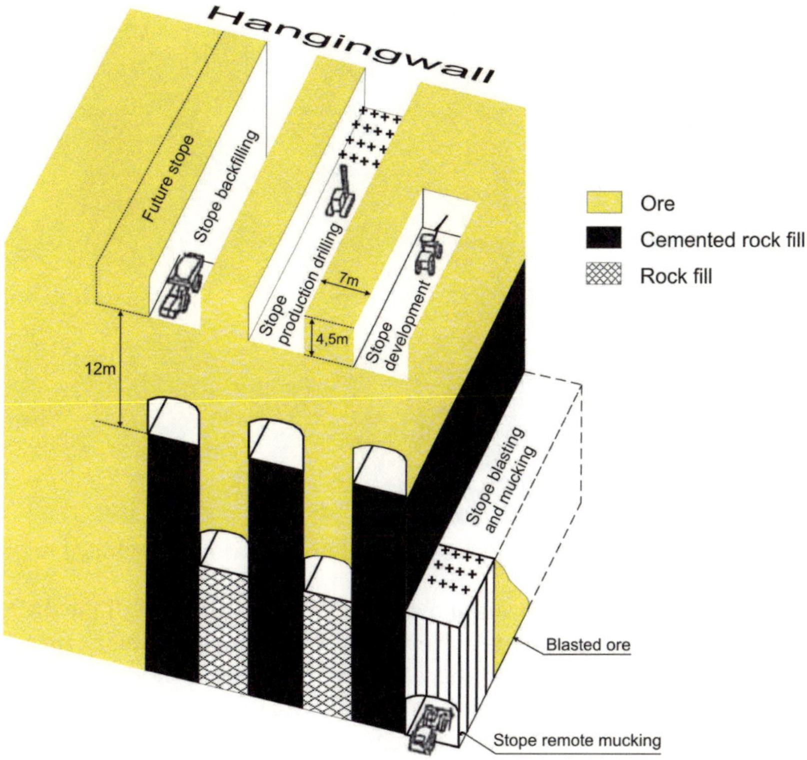

Fig. 4. Development, production and backfilling operations.

	Ore	Basalt	%5 Cemented Backfill
Density (gr/cm^3)	4,1	2,7	2,1
Uniaxial Compressive Strength (MPa)	55	65	6,5
Indirect Tensile Strength (MPa)	4,9	6,1	0,7
Cohesion (C) (MPa)	12,3	13,5	1,4
Internal Friction Angle (ɸ)	43,8	45,5	25,4
Young Modulus (MPa)	33500	47600	2850
Poisson's Ratio	0,26	0,25	0,34

Table 1. Geomechanical properties of the ore, basalt and backfill.

When the stresses are analysed in the case in which the stopes are extracted moving upward and the cavities were filled with either the cement added backfill or ordinary uncemented backfill, it was understood that no stability issue would arise. An example of vertical stress distribution around open stopes and pillars is shown in Figure 5 [5].

Although the production sequence of the stopes proceeds from bottom to the top, not all the stopes in the lower level are extracted to move on to the production of the upper levels. There are two main reasons for this condition. With respect to the former one; since the head entry of the lower stope is employed again as the tail entry of the upper stope later, it can not be filled with any backfill material, which means that the bottom part of the tail entries

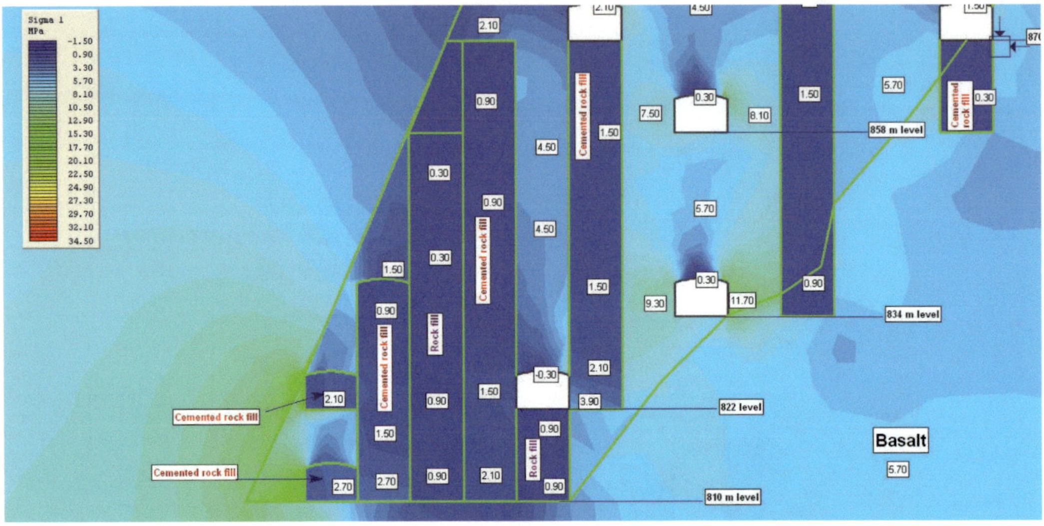

Fig. 5. Vertical stresses between 810 m level and 858 m level at west sector of mine.

of the upper levels is composed of backfill material. In this case when all the drifts come side by side (as the width of the drift equals the width of the stope), stability problems will be observed as there will be no support beneath the prospective unextracted stopes. With respect to the latter one; since there are many stopes at the same level, the production of many stopes from the same level entries will be hard.

Therefore, to make simultaneous productions out of a couple of levels will be convenient from the aspects of stope stability, machinery-equipment organization and the ore grade optimization. Generally, the production sequence from bottom to the top happens to be in triangular or diagonal shape. This mentioned production sequence is very important from the view of stability of the underground mine.

5. References

[1] Köse, H., Yalçın, E., Gönen, A., Evaluation of Etibakır Asikoy underground copper mine the second-stage project tecnically (in Turkish), Project report, Dokuz Eylul University Foundation, Izmir, 2007.

[2] Koç, Ş., Unsal, A., Kadıoğlu, Y. K., Geological, Geochemical and Geotechnical Properties of Küre(Kastamonu) Ore Deposit, (in Turkish), MTA Journal 117, pp 41-54. 1995.

[3] Kuşçu, İ., Erler, A., Pyrite deformation textures in the deposits of the küre mining district (Kastamonu-Turkey), Turkish Journal of Earth Sciences (Turkish J. Earth Sci.), Vol. 11, pp. 205-215, 2002.

[4] Gönen, A., Investigating The Underground Mining Method Of Etibakır Co. Küre Copper Mine Technically And Economically. (in Turkish), Phd Thesis, Dokuz Eylul University The Graduate School of Natural and Applied Sciences, İzmir, 2010.

[5] Köse, H., Yalçın, E., Gönen, A., Investigation of Stability of Open Stopes and Pillars in Longhole Stoping Method at the Asikoy Underground Copper Mine (in Turkish), Project report, Dokuz Eylul University Foundation, Izmir, 2007.

5

Leak Tightness of Underground Carbon Dioxide Storage Sites and Safety of Underground CO_2 Storage by Example of the Upper Silesian Coal Basin (Poland)

Eleonora Solik-Heliasz
Central Mining Institute
Poland

1. Introduction

The region of the Upper Silesian Coal Basin (USCB, southern Poland) was a subject of investigations referring to possibilities of underground storage of carbon dioxide. It is an industrial region characterized by high density of industrial infrastructure on the surface of the ground and underground as well as considerable emission of CO_2 into the atmosphere. Among significant objects one should mention the presence of 53 mines of hard coal (Fig. 1). In these mines, there is or was conducted in previous years exploitation of coal deposit to the depth of about 1000 m below the terrain surface level. Location of underground storage sites within a small distance from an industrial emitter of CO_2 might positively influence the cost-effectiveness of underground storage - if only by reducing costs of gas transport. However, with the current infrastructure, a selection of a proper site for storage necessitated conducting additional investigations, mainly detailed, multi-stage analysis of CO_2 injection safety. Below, there are presented outcomes obtained in the years 2007-2010 in the framework of the research project the Technological Initiative I, entitled: "*Study of safe carbon dioxide storage by example of the Silesian agglomeration*", and dedicated researches carried out because of commissions received from concerned economic entities.

2. Selection of CO_2 storage location

Selection of suitable location of underground storage is a prerequisite for successful process of CCS (*carbon capture storage*). Prospective site for storage must be market out by advantageous reservoir parameters, which was earlier mentioned by numerous researchers (Bachu et al., 2007; Bruining et al., 2004; Chadwick et al., 2008; Kumar et el., 2005; Obdam et al., 2003; Solik-Heliasz, 2010a). Both, safety of storage and effectiveness of CO_2 injection must be also ensured. In the course of conducted examinations it was found out that in the region of USCB, underground storage sites must be located only beyond areas of active and liquidated mines but also outside areas of urban agglomeration and huge industrial plants. It is so because it has been ascertained that areas of underground mining are potentially seismic regions. Rock mass parted with mine workings can be a source of vibrations, which in unfavourable hydrogeological conditions might activate new paths of underground water migration as well as injected gas and lead to their unforeseen dislocation. Also areas of cities and large plants have been excluded from CO_2 storage - even at the deepest depths. In this case it does not, however, stem from some real threat but from a lack of sense of security which can concern the local community. In the investigated region it has also turned out to be crucial that prospective storage sites do not influence other undertakings of utilitarian character existing and planned in the vicinity of their borders. It can refer to both, workings of underground mines as well as water intakes, geothermal boreholes etc. Owing to the importance of the problem in question it became a subject of detailed investigations.

Having in mind the selected factors, the sandy and water-bearing horizon of the Dębowiec layers in the region of Skoczów-Zebrzydowice (Fig. 1) has been indicated for the injection of carbon dioxide. This horizon occurs in the forefield of USCB, beyond mines area, in overburden of coal series. Additionally, initially for the injection were qualified hard coal seams no. 405 and 510, deposited in the area of reserve mine field Pawłowice, at the depth of 1100-1500 m, as well as workings of hard coal mines Krupiński, Silesia and Brzeszcze - after their previous liquidation (Solik-Heliasz, 2010b).

The issue of safety of CO_2 injection was considered with reference to the area of the Skoczów-Zebrzydowice storage site. Nevertheless, the addressed problems can also refer to the remaining storage sites.

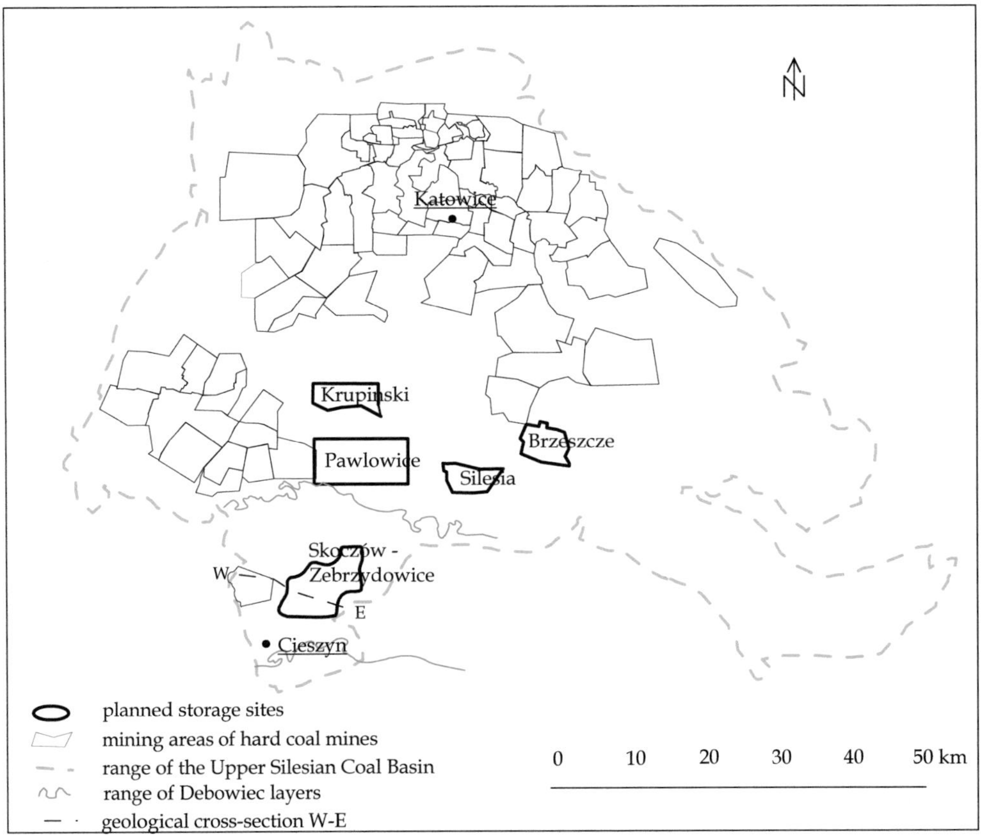

Fig. 1. Location of planned carbon dioxide storage sites in the region of USCB (Poland).

3. Examination of storage site leak tightness

Rock formations of the Skoczów-Zebrzydowice storage site are connected with Neogenic rock formations of the Dębowiec layers. They occur within the limits of regional structure of the Carpathian Foredeep, which stretches at the length of over 200 km along the southern border of Poland. The layers, deposited on morphologically diversified surface of carbon roof, are covered with thick complex of the Miocene claystones (Fig. 2). The existing hydrogeological conditions as well as parameters of reservoir rock formations, which are presented in Table 1, create favourable conditions for CO_2 storage. Nevertheless, the decisive role in the final selection of a site for injection played the analysis of storage site leak tightness and its potential influence on other underground objects in its vicinity. Leak tightness has been documented on the basis of geological cartography methods (maps, cross-sections), results of laboratory and *in situ* tests on hydro-geological and strength parameters of rocks, analyses of water chemical mechanism, results of geophysical tests and modelling.

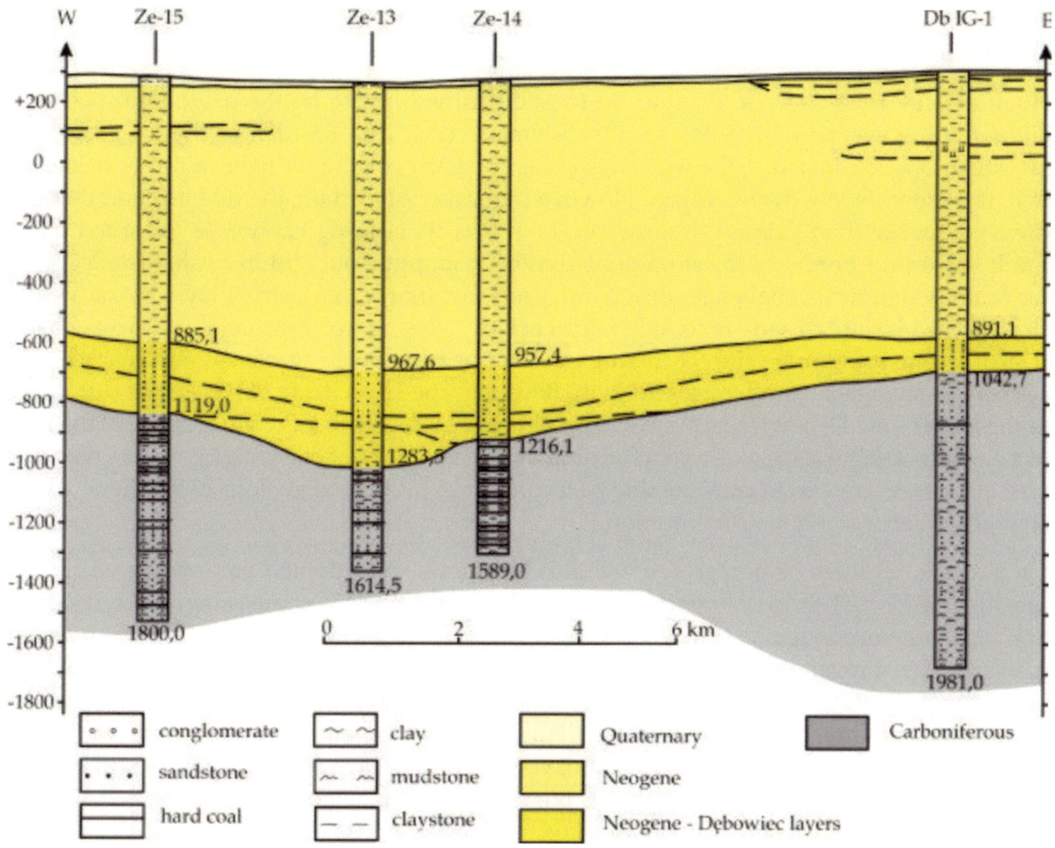

Fig. 2. Geological cross-section of the region of the Skoczów-Zebrzydowice storage site.

Rock parameters	Parameters values
Porosity of the Dębowiec layers rock formations [%]	8-15
Permeability of the Dębowiec layers rock formations [mD]	13-103
Compressive strength / tensile strength, R_c/R_r [MPa]: - rocks in overlay of the storage site	5.0-39.9/-
- storage site rock formations	4.0-56.0 / 0.1-2.4
- rocks in storage floor	26.9-44.7/1.8-3.8
Thickness of storage rock formations [m]	70-250
Thickness of isolation rock formations in storage roof [m]	720-1030
Effective CO_2 storage capacity [$Mg \cdot 10^6$]	24.1

Table 1. Values of selected parameters of the Dębowiec layers in the area of the Skoczów-Zebrzydowice storage site.

The devised static model of the Dębowiec layers has shown that clay slates occurring in the overburden and in the floor of CO_2 reservoir are impenetrable. Leak tightness was also confirmed by chemical analyses of waters, their hydro-chemical indicators and isotopic composition of oxygen, $\delta^{18}O$. At the horizon of the Dębowiec layers, paleoinfiltrational waters have been found of Cl-Na and Cl-Na-Ca type, and mineralization up to 98 g/dm^3. They indicate lack of connection between the hydraulic reservoir with the terrain surface.

Hindrance in a selection of storage site location turned out to be the neighbourhood of a large fault zone with 400-600 m displacement (Fig. 3). Results of hydro-geological investigations conducted in the workings of nearby hard coal mines have admittedly shown that this zone is not water filled. However, in case of carbon dioxide injection under pressure higher than primary hydrostatic pressure, its clearing cannot be excluded. As a result of this, the border of the storage site has been mapped out within a safe distance from the fault. A significant challenge turned out also a fact that water-bearing layer in the area of storage yard is not "closed" in its side part. Lack of "closing up" of storage site is a problem referring to many storage sites occurring in water-bearing horizons of regional character. In case of the storage site under examination, however, the knowledge of the height difference of the floor of the Dębowiec layers was applied. The storage site has been located in the area of two floor structural dips of layers forming local valleys about 300 m deep. In this way, the area of storage site was in considerable part contoured by natural geological frontiers, which should influence safety of CO_2 injection.

On the leak tightness of storage site will also have influence adopted parameters of carbon dioxide injection. A threat can particularly result from the adopted pressure of CO_2 injection. Too high pressure might lead to fracturing of reservoir rocks and/or neighbouring rocks. It is even more dangerous as rock formations of the Dębowiec layers display not too high values of strength parameters (Table 1). Fracturing of reservoir rocks is generally treated as a factor posing threat for storage safety. However – an opinion may be expressed here – that for reservoir rocks of medium and average values of hydro-geological parameters, controlled fracturing, limiting injection boreholes to the region can improve absorbing power of reservoir rock formations. In dynamic models of CO_2 propagation in the horizon of the Dębowiec layers, developed with the use of numerical programme TOUGH-2 (Pruess, 1991), effects of injection were analysed: 100 000, 300 000 and 500 000 Mg CO_2/year during the period of 30 years. As a result of CO_2 injection a repression cone will arise. It has

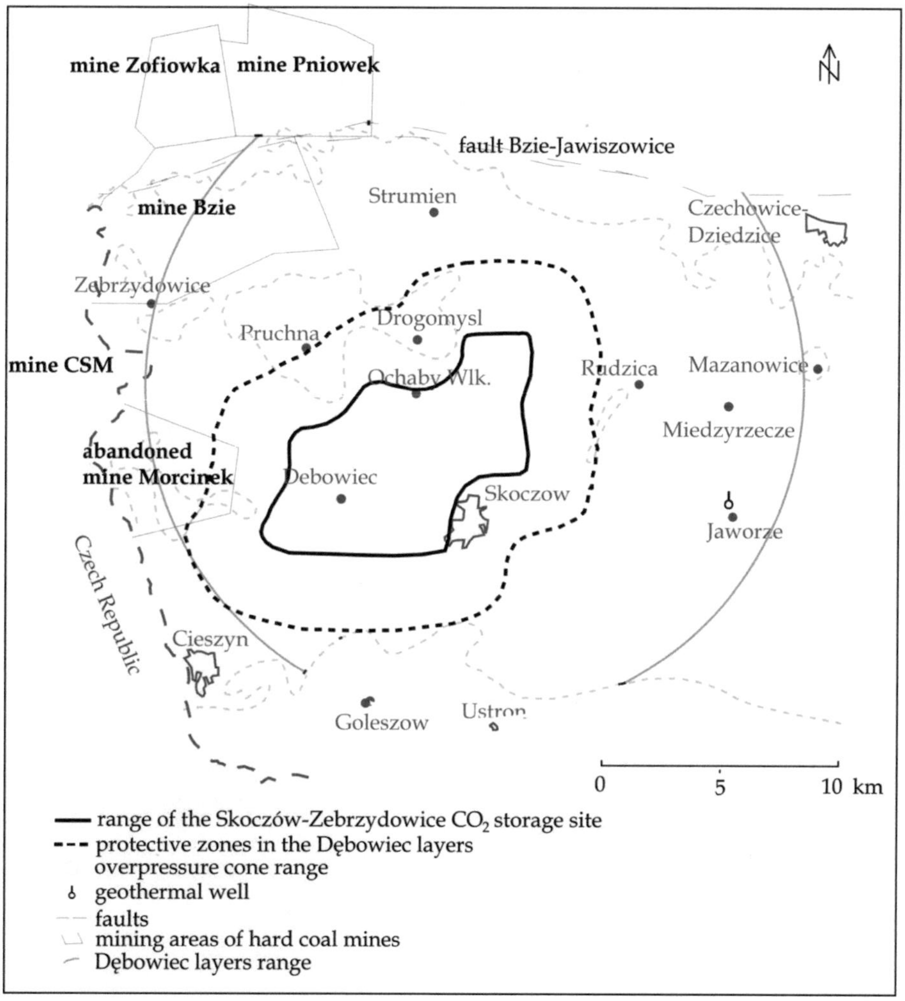

Fig. 3. Location of the Skoczów-Zebrzydowice storage site and major environmental features.

been ascertained that the optimum amount which can be injected through two injection boreholes amounts to 2x150 000 = 300 000 Mg/year. Injection of CO_2 into the floor part of the Dębowiec layers (Fig. 4a) will cause rapid increase of the pressure of water-gas medium in both holes from the initial value of 10.75 MPa to 12.6 MPa. During a short period of time maximum pressure in the region of injection boreholes can be similar to the pressure of rock fracturing. However, in the roof part of the Dębowiec layers the increase of pressure will mark itself much less clearly (Fig. 4b). On the other hand, saturation of water-bearing horizon with carbon dioxide will demonstrate unlike dynamics. During carbon dioxide injection (period of 0-30 years) CO_2 will be mainly accumulated in floor part of the Dębowiec layers (Fig. 5a). In the later period (period of 30-100 years), it will fill the space of the roof part (Fig. 5b). Results of analytical calculations confirmed that after 5 years of CO_2 injection, the radius of formed repression cone will amount to 11.8 km and encompass in its range both, storage site area as well as an area adjacent to it (Fig. 3).

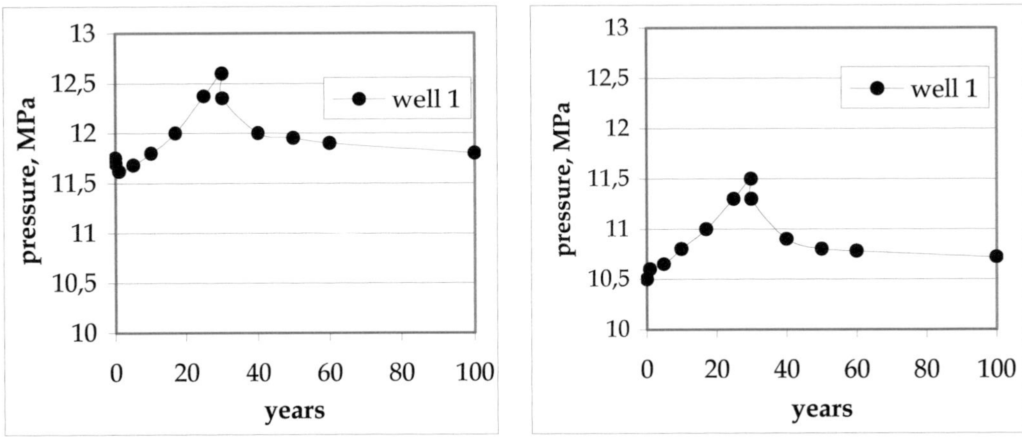

(a) in floor part (b) in roof part

Fig. 4a,b. Pressure changes of water in the Dębowiec layers as a result of CO_2 injection in the amount of 300 000 Mg/year (according to Solik-Heliasz, on the basis of data obtained from Chećko, 2010).

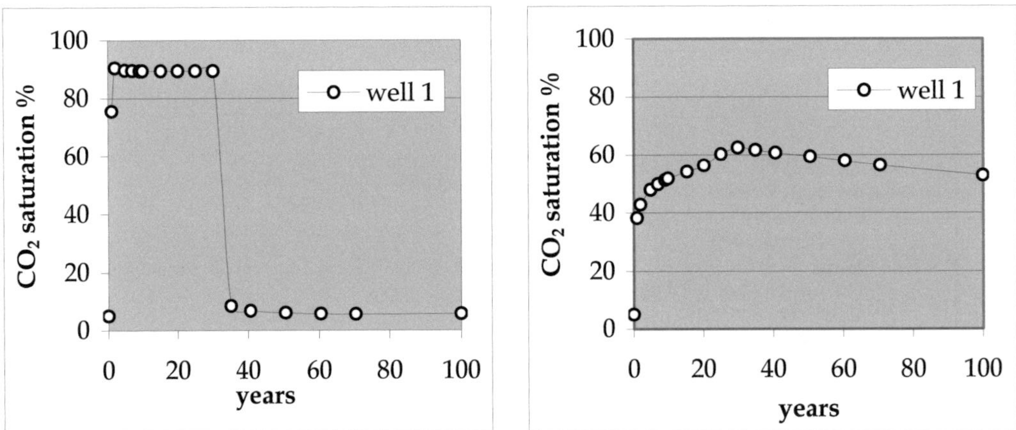

(a) in floor part (b) in roof part

Fig. 5a,b. Distribution of CO_2 saturation in the Dębowiec layers (according to Solik-Heliasz, on the basis of data obtained from Chećko, 2010).

4. Analysis of storage site impact

Hitherto in Europe and in the world, locations of carbon dioxide storage sites are planned mainly in water-bearing horizons not utilized economically. This can however be subject to change due to building new underground objects. Then, a problem may arise connected with their interaction and a necessity to determine a safe distance between them. This problem has especially occurred in the examined region of USCB, especially with reference to workings of hard coal mines. From the conducted research works it follows that the approval of storage sites locations in industrialized regions or regions adjacent to them, requires carrying out a detailed analysis of possible interactions. A two-stage analysis was performed. In the initial stage it covered determination of possible impact of storage site on other present and planned underground objects in its vicinity. In the subsequent stage an extreme scenario, in which consequences of possible penetration of CO_2 in the direction of other underground objects were analysed.

With reference to the Skoczów-Zebrzydowice storage site, it was examined how the storage site influences the following (Fig. 3):

1. planned geothermal water intake in the Dębowiec layers;
2. water-bearing horizon of the Dębowiec layers in the overburden of mines Morcinek and Bzie.

These objects are situated beyond the storage site area, thus a source of threat can come mainly from overpressure created in the horizon of the Dębowiec layers. In connection with this, it turned out justifiable to provide the answer to possible physical effects of overpressure. The results of preliminary analyses of simulations have shown that it can be (apart from rock formations fracturing mentioned before) the increase of water flow velocity in the horizon of the Dębowiec layers. With reference to the planned geothermal waters intake - the speed increase can cause increased water inflow to the intake and at the same time improve its effectiveness. On the other hand, with reference to the planned mine Bzie - in case of drainage of the Dębowiec layers in its area, which is to serve the increase of security of conducting mining exploitation - the increase of water flow might lead to increase of intakes to future drainage wellbores and as a result influence sooner attainment of the assumed depression of water level. The drainage process must be accurately monitored. It has been estimated that because of security reasons, arising cone of depression must not cross a border of protection zones established around the storage site (it will be developed in the further part of the work). In the extreme scenario, in contrast, the following were analysed:

1. results of possible penetration of CO_2 from the storage site to lower deposited workings in liquidated mine Morcinek;
2. threat to the stability of protecting pillar existing between workings of Polish mine Morcinek as well as active Czech mine ČSM.

Poor water leading of rock formations in the floor of storage site confirmed the results of *in situ* observations carried out during the functioning of mine Morcinek. Moreover, the dynamic model of the storage site area has not confirmed a possibility of carbon dioxide penetration into the foundation as well as its horizontal migration in the direction of mine workings. Despite that, the obtained results were treated with due caution. The numerical model requires designing a scheme of boundary conditions, whereas rock formations of storage site foundation constitute numerous interbedding of clay slates, mudstones and subordinately sandstones with coal seams. In spite of appropriate recognition of their parameters - general assessment of conductivity of this part is surely not sufficient. Thus, it turned out necessary to utilize the results of mining recognition as well as calculation methods applied in this field.

Workings of coal mine Morcinek are deposited at the depth of 750-1100 m below the terrain surface level. They are submerged and form huge reservoirs of underground waters. Their water capacity has been determined on the basis of calculation method by Rogoż (Rogoż,

2007) to be in total 2,3 $\cdot 10^6$ m^3 (Solik-Heliasz, ed., 2009). What is more, in side walls of a long network of mine workings: longwall, heading and drifts, occur remainders of hard coal deposits which can even amount to several dozen million tonnes. Results of laboratory analyses have shown that hard coal has substantial sorption capacity of carbon dioxide (Ceglarska-Stefańska et al., 2008). Table 2 presents CO_2 storage capacity in workings of selected mines, Krupiński and Silesia, in case of creating low-pressure reservoirs in them, of pressure $p<0.6$ MPa. These mines are situated in comparable geological conditions to mine Morcinek.

Parameters	Silesia Mine	Krupiński Mine
Volume of mine workings: longwall, passageway and caving fractures [$m^3 \cdot 10^6$]	25.08	5.80
Mass of left hard coal [$Mg \cdot 10^6$]	78.75	43.56
Total amount of CO_2 possible to be stored in mine workings and in coal reminders [$Mg \cdot 10^6$]	2.9-6.7	0.9-1.9

Table 2. CO_2 storage capacity in workings and coal reminders in mines Silesia and Krupiński in GZW (Solik-Heliasz, 2010b).

The presented results show that workings of mine Morcinek are capable to absorb substantial amount of CO_2 in case of its potential migration from the Skoczów-Zebrzydowice storage site. Separate issue is a threat which can create the increase of water-gas medium in mine workings of mine Morcinek, on the stability of protecting pillar separating workings of this mine from workings of active mine ČSM (Fig. 3). The stability of the pillar has been verified on the basis of calculation methods utilized in underground mining. It turned out helpful to adopt i.a. the *Slesariew*'s formula (Frolik, 1998), enabling to determine a safe pillar width:

$$D_{min} = g\sqrt{60p} \quad (1)$$

where: D_{min} – minimum width of protecting pillar, m

g - average thickness of reservoir series, m

p - target pressure in the reservoir, MPa.

Increase of pressure in abandoned working of mine Morcinek even by 50% in comparison to the primary hydrostatic pressure amounting to 7.8 MPa, will not threaten the stability of the existing pillar. Its present width which is 100 m will be sufficient.

Conducted examinations have shown that the process of the storage site exploitation does not encounter difficulties which could lead to its premature closure.

5. Will carbon dioxide storage sites require protection?

In EU Directive concerning underground storage of carbon dioxide (Directive, 2009) it is recommended that underground storage ought to be safe and injected carbon dioxide permanently bound with reservoir rock formations in the period of several hundred years and more. On the basis of its guidelines as well as experiences gained so far (i.a. Chadwick et.al., 2008, Frolik et al., 2006), the ranges of storage sites in the area of USCB were established. Does it however mean that in the direct vicinity of storage sites - which theoretically should be safe - other utilitarian enterprises, e.g. water intakes, storage yards etc. can be located with complete confidence?

The rock formations of the Dębowiec layers mentioned above are characterised by changeability of hydro-geological parameters. Similar Carboniferous rock formations deposited in their floor, which additionally demonstrate high lithological diversification. Rocks of these two geological series are a subject of research conducted since the end of the 60's of the 20th century. Despite that, the diversification, which has been mentioned, can become a potential source of threat for safe storage of CO_2. Having that in consideration, it has been proposed to establish protection zones around CO_2 storage sites in USCB. They would constitute additional protection of storage sites against the influence of other utilitarian enterprises as well as other underground objects against storage site impact. Cap rock is a typical example of protection zones. Nonetheless, zones of this type should be also established in the side regions of the storage site and in its floor. The concept of protection zones location in various types of CO_2 storage sites are presented in Figures 6a-c.

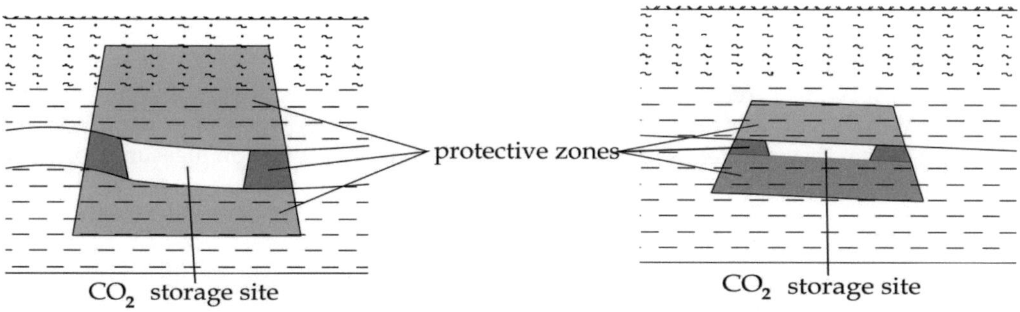

(a) sites created in water-bearing horizons

(b) created in hard coal seams

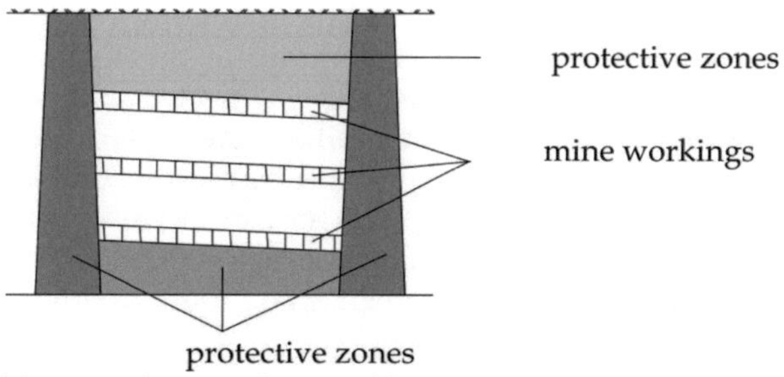

(c) created in workings of liquidated hard coal mines

Fig. 6a-c. Concept of location of protection zones for carbon dioxide storage sites.

Methodology of delineation of protection zones width ought to become a subject of further research. Preliminary calculations made on the basis of the formula (1), with the assumption that CO_2 carrier after its injection into reservoir rock formations will be the flow of underground waters, have shown that the width of protection zones can be diversified. Appropriate values are presented in Table 3.

Location of protection zones for CO_2 storage sites situated in:	Minimum width of protection zones in reservoir rock formations	Notice
	m	
Water-bearing horizon - in the region of CO_2 storage site Skoczów-Zebrzydowice	2540-2920	High-pressure CO_2 reservoir; $p \geq 10.75$ *MPa*
Hard coal seams 405 and 510 - in the area of CO_2 storage site Pawłowice	147-196	High-pressure CO_2 reservoir; $p=10$ *MPa*
Workings of hard coal mines Krupiński and Silesia - after their liquidation	1800-2400	Low-pressure CO_2 reservoirs; $p< 0.6$ *MPa*

Table 3. Minimum width of protection zones around underground carbon dioxide storage sites in the area of USCB (Poland).

Delineation of protection zones around carbon dioxide storage sites can have, however, economical aspect, as it requires exclusion of additional parts of rock mass from economic activity. Nevertheless, the growing number of utilitarian enterprises in water filled rock series causes that they may require introduction of additional protection measures for them.

6. Conclusions

There is a possibility to create carbon dioxide storage site in the region adjacent to highly industrialised area of USCB. However, approval of its location took place only after conducting tests on storage site leak tightness and two-stage analyses of its potential impact on other underground objects located in the vicinity (workings of hard coal mines, water intakes). For the assessment of carbon dioxide storage safety, results of model research have been used which were supplemented with results of mining recognition (hydrodynamic regime in the area of the surrounding mine workings of the abandoned hard coal mine) using the calculation methods utilized in this field (connected with water capacity of mine workings, minimum width of the protecting pillars and their stability). The Skoczów-Zebrzydowice storage site will not negatively influence other underground objects in its vicinity. However, due to the fact that their number can increase in the future, it has been proposed to establish protection zone around storage site. It will constitute additional protection and determine intransgressible influence boundary of other underground objects planned to be built nearby.

7. References

Bachu, S., Bonijoly, J., Burruss, R., Christensen, N.P., Holloway, S. & Mathiassen, O.M. (2007). Estimation of CO_2 Storage Capacity in Geological Media - Phase 2, *Prepared* *by the Task Force on CO_2 Storage Estimation for the Technical Group (TG) of the Carbon sequestration Leaderskie Forum,* February 12, version 1.5. URL: http://www.cslforum.org/documents/PhaseIII/ReportStorageCapacityMeasurementTask, Force.pdf.

Bruining H., Bossie-Codreanu, D., Busch, A., Choi, X., de Schmedt, G., Frobel, J., Gale, J., Grabowski, D., Hadro, J., Hurtevent, D., Jura, B., Kretchmar, H., Krooss, B., Krzystolik, P., Mazumder, S., Muller, G., Pagnier, H., Reeves, S., Skiba, J., Stevens, S., van Bergen, F., van der Meer, B., Wenting, P., Winthaegen, P. & Wolf, K. (2004). Recopol - field experiment of ECBM-CO_2 in the Upper Silesian Basin of Poland, *7th International Conference on Greenhouse Gas Control Technologies,* Vancouver, Canada.

Ceglarska-Stefańska, G., Zarębska, K. & Wolszczak, J. (2008). Sorption of pure components and mixtures CO_2 and CH_4 on hard coals, *Mineral Resources Management* Vol. 24(4/1), pp.123-131.

Chadwick, A., Arts, R., Bernstone, C., May, F., Thibeau, S. & Zweigel, P. (2008). *Best practice for the storage of CO_2 in saline aquifers. Observations and guidelines from the SACS and CO_2STORE projects,* Keyworth, Nottingham British Geological Survey.

Chećko, J. (2010). Modelling of underground reservoir for CO_2 storage in brine-bearing formations, *in* Wachowicz, J. (ed.), *Study of safe carbon dioxide storage by exemple of Silesian agglomeration,* Wyd. GIG, Katowice, pp. 52-75 (in Polish).

Dyrektywa Unii Europejskiej i Rady 2009/31/*WE z dnia 23.04.2009,* dotycząca podziemnego składowania CO_2.

Frolik, A. (1998). Ocena szczelności przeciwwodnych filarów bezpieczeństwa. *Prace GIG.* VII Konferencja: Problemy geologii w ekologii i górnictwie podziemnym, Szczyrk, Poland, pp. 231-240.

Frolik, A., Kidybiński, A., Marcak, H., Nierobisz, A., Rychlicki, S. & Solik-Heliasz, E. (2006). Assessment methods of possibilities of hydrocarbons storage in underground mine workings, *in* Kidybiński, A., Siemek, J. (eds.), *Underground repositories of gas in abandoned hard coal mines,* Wyd. GIG, Katowice, pp. 59-95 (in Polish).

Kumar, A., Noh, M.H., Sepehrnoori, K., Bryant, S.L. & Lake, L.W. (2005). *Simulating CO_2 storage in deep saline aquifers, in Carbon Dioxide Capture for Storage in Deep Geologic Formations – Results from the CO_2 Capture Project,* Elsevier, London Vol. 2.

Obdam, A., van der Meer, L.G.H., Kervevac, C., Bech, N. & Wildenborg, A. (2003). Effective CO_2 storage capacity in aquifers, gas fields, oil fields, and coal fields, *Proceedings of the 6th International Conference on Greenhouse Gas Control Technologies,* Gale and Y. Kaya, 1-4, October 2002, Kyoto, Japan, Pergamon.

Pruess, K. (1991). TOUGH2 - A General-Purpose Numerical Simulator for Multiphase Fluid and Heat Flow. *Earth Science Division,* Lawrence Berkeley Laboratory, University of California, Berkeley, CA 94720, Report LBL-29400.

Rogoż, M. (2007). *Dynamika wód podziemnych,* Wyd. GIG, Katowice.

Solik-Heliasz, E. (ed.) (2009). *Atlas of geothermal energy resources in the Upper Silesian Region – the Neogene, Carboniferous and Devonian formations,* Wyd. GIG, Katowice (in Polish).

Solik-Heliasz, E. (2010a). Development of potential capacity of CO_2 storage in deeply located brine-bearing formations in the region of Silesian agglomeration, *in* Wachowicz, J.

(ed.), *Study of safe carbon dioxide storage by exemple of Silesian agglomeration*, Wyd. GIG, Katowice, pp. 15-51 (in Polish).

Solik-Heliasz, E. (2010b). Possibilities of underground CO_2 storage in the Upper Silesian region, *Mineral Resources Management* Vol. 26 (3), pp. 53-65.